T0137079

Genetic and Evolutionary Computation

Series Editors

More information about this series at https://link.springer.com/bookseries/7373

Wolfgang Banzhaf · Leonardo Trujillo ·
Stephan Winkler · Bill Worzel
Editors

Genetic Programming Theory and Practice XVIII

Editors
Wolfgang Banzhaf
Department of Computer Science
and Engineering
Michigan State University
East Lansing, MI, USA

Stephan Winkler
School of Informatics, Communications
and Media
University of Applied Sciences Upper
Austria
Hagenberg, Austria

Leonardo Trujillo
Tecnológico Nacional de México/IT de
Tijuana
Tijuana, Baja California, Mexico

Bill Worzel
Evolution Enterprise
Ann Arbor, MI, USA

ISSN 1932-0167 ISSN 1932-0175 (electronic)
Genetic and Evolutionary Computation
ISBN 978-981-16-8115-8 ISBN 978-981-16-8113-4 (eBook)
https://doi.org/10.1007/978-981-16-8113-4

This Springer imprint is published by the registered company Springer Nature Singapore Pte Ltd.
The registered company address is: 152 Beach Road, #21-01/04 Gateway East, Singapore 189721, Singapore

Foreword

This book highlights the extraordinary recent progress in genetic programming. I wonder what John Holland would have made of it. Many people imagined computers that could program themselves, but John took seriously the idea of using selection to evolve working programs. The field of genetic programming has itself now evolved, as dozens of programmers set a dizzying array of strategies competing to shape effective efficient programs as quickly as possible. The results are astounding.

I thought I understood natural selection, but then I discovered Adaptation in Natural and Artificial Systems in 1976 on the front table in the original Borders bookstore in Ann Arbor. John Holland's book made me realize that the principle of selection is much more general, and that selection in silico can reveal a lot about natural selection. Reading this current volume brings this full circle and at a much higher level. It describes the strategies genetic programmers use to overcome the obstacles that limit what natural selection can do. In particular, readers will discover new strategies for optimizing levels of variation, defining goals, and structuring selection in ways that transcend and illuminate the limits of natural selection.

Optimizing mutation rates is a core task for genetic programmers; they have a free hand to set mutation rates and their timing ("rampant mutation") to maximize the speed of evolution and minimize getting stuck on suboptimal peaks. They can measure diversity in new ways ("phylogenetic diversity"). Natural selection is limited by comparison. Mutator genes that might otherwise benefit the species are selected against, although some bacteria increase mutation rates temporarily when under stress.

The process of defining goals is also very different. For genetic programs, that means reaching a reliable and efficient solution to a defined problem as quickly as possible. Each agent is evaluated by its contribution to the defined goal, and multiple goals are possible. Natural selection has no goal; it just mindlessly increases the prevalence of whatever variants result in individuals who reproduce more than other individuals. So variants that decrease health, happiness, longevity, or cooperation quickly become more common if they increase Darwinian fitness.

Strategies for selecting superior programs are at the core of genetic programming. Lexicase selection is superior to tournament selection, especially in its ability to explore the full fitness space. The value of offspring can even be predicted by data about the parents. Natural selection, by contrast, mindlessly increases and decreases the prevalence of genetic variants in proportion to their effects on Darwinian fitness, in competition with stochastic influences that can eliminate useful mutations or take deleterious ones to fixation.

The problem of getting stuck at suboptimal peaks is faced by both genetic programming and natural selection, but the options are much more diverse for programmers. Natural selection cannot start fresh. Path dependence restricts it to small changes that leave many traits suboptimal, such as the shared passageway for air and food, and blood vessels that run between light and the retina.

I came to the workshop and this book looking for examples of how genetic programs fail that might illuminate the evolutionary reasons why we are vulnerable to disease. But from what I can tell, genetic programs are not vulnerable to failures akin to cancer, heart attacks, diabetes, or autoimmune disease. The intrinsic advantages of genetic programs described above offer only a partial explanation. Instead, I think the vulnerabilities and the robustness of organic systems result from billions of years of selection that have created organic complexity that defies simple description. Our human minds were shaped to seek simplicity, however, so we tend to view bodies as if they were designed, with discrete parts each with a specific function. The prevalence of this tendency to "tacit creationism" is a major obstacle to full biological understanding. Deep learning neural networks also can be similarly impossible to comprehend, but there is hope that genetic algorithms will make them less opaque.

I come away impressed by the vibrant community of scholars and scientists who are using genetic programming to create programs very different from the products of natural selection. They leave me astounded that natural selection has done so well despite its limits. I hope someone who knows more than I do about genetic programming and natural selection will compare and contrast them in much more detail in order to illuminate them both.

Ann Arbor, MI, USA
October 2021

Randolph M. Nesse
http://nesse.us

Preface

The preparations for the eighteenth edition of the workshop on Genetic Programming Theory and Practice (GPTP) began back in the autumn of 2019. Some months before, the seventeenth edition of GPTP had been a great success: After 16 annual events in Ann Arbor, at the University of Michigan, it had been the first edition of GPTP to be held at Michigan State University in East Lansing. For the eighteenth edition, GPTP 2020, we planned to go back to Ann Arbor. So, similar to the years before, we started to organize venue details, invited potential speakers and workshop participants, and asked sponsors if they would be willing and able to support the event. By February 2020, everything was quite well on track, and some of us already had planned their travel to GPTP 2020.

But then the COVID-19 pandemic struck the entire world. By March 2020 it was clear that an in-person event in Michigan would not be possible in May 2020, and with a very heavy heart, we had to cancel the event for 2020. It was the first time since 2003 that no GPTP took place.

As time went on, in autumn 2020 we started discussions about organizing the 2021 GPTP workshop. In the beginning we were optimistic that a normal event could be possible, but it quickly dawned on us that this would be highly improbable. It was clear that universities would still be shut in the spring of 2021 and that traveling would not be safe by then either. But should we cancel GPTP once again? We were all relatively used to online meetings—but could a workshop such as GPTP "work" in an online form? Considering that the real great moments and discussions at GPTP happen in the discussion sessions and after the talks, when the ideas come up discussing the day's sessions at a drink in the evening, what should we do?

As the GPTP workshop has become one of the most important meeting points for our community, where people meet and can share their thoughts freely and get feedback for their ideas should we really cancel once again? Especially in the challenging times of a pandemic, we felt that it would even be more important to keep the community connected and alive!

So, after several weeks of discussions, we decided to give it a try—GPTP 2021 would be an online event! Again, we contacted potential speakers, participants, and sponsors, keeping in mind that for an online version of the workshop it would be very important not to make the event too big. The feedback from people we approached was very positive—almost all of them accepted their invitations.

And so, May 2021 came along, and GPTP 2021 was held as an online event via Zoom—and it was great! The talks were interesting, the discussions were very intense and, similar to the in-person events, many new ideas were presented and created during the event.

Each day there was a keynote talk, followed by shorter presentations and discussions as well as a special discussion sessions about the topic of the keynote. On Day 1, Elizabeth Barnes from the Department of Atmospheric Science at Colorado State University gave a keynote about viewing anthropogenic change through an AI lens. Climate change is a global problem and a threat to us all, and there is a lot of problems where computer scientists, and especially understandable AI models created by genetic programming could help.

On Day 2, Randolph Nesse from the Center for Evolution and Medicine at Arizona State University gave a keynote about evolutionary medicine, why evolved systems fail, and the mystery of health. This talk was perfectly aligned with the tradition of GPTP, as we always strive to combine computer science, evolutionary algorithms, and interdisciplinary thinking.

Finally on Day 3, David Andre from Google-X talked about pitfalls and things that might go wrong when deploying GP systems, especially in the field of finance—his talk was entitled provocatively "GP considered Dangerous". Again, this keynote sparked many intense discussions as those are topics we all can relate to.

The collection you hold in hand contains the written final contributions submitted by the 18th workshop's participants. Each contribution was drafted, read, and reviewed by other participants prior to the workshop.

We are very glad that we were able to carry on with the spirit of GPTP in 2021, as a special place in the genetic programming community in an unusually intimate, interdisciplinary, and constructive atmosphere. It brings together researchers and practitioners who are eager to engage with one another in thoughtful and unhurried discussions of the major challenges and opportunities in our field.

Acknowledgements

We would like to thank all of the participants for making GP Theory and Practice a successful workshop 2021. As is always the case, it produced a lot of interesting and high-energy discussions, as well as speculative thoughts and new ideas for further work. The keynote speakers delivered thought-provoking talks from diverse perspectives.

We would also like to thank our financial supporters for making the existence of GP Theory and Practice possible for now nearly two decades. For 2021, we are grateful to the following sponsors:

- John Koza

- Gilda Cabral and Michael Korns
- Mark Kotanchek at Evolved Analytics
- Stuart Card
- Michael Affenzeller at the University of Applied Science Upper Austria

A number of people made key contributions to the organization of the workshop. Foremost among them is Constance James, who helped behind the scenes before, during, and after the workshop. Special thanks to Michigan State University, for providing Zoom online services, as well as to the Springer-Nature Publishing Company, for producing this book. We are particularly grateful for contractual assistance by Mio Sugino, Springer-Nature Tokyo, and Ronan Nugent, Springer-Nature Heidelberg.

We would also like to express our gratitude to Carl Simon at the Center for the Study of Complex Systems at the University of Michigan and to Erik Goodman and Charles Ofria at the BEACON Center for the Study of Evolution in Action at Michigan State University for their continued support. Free online social space was provided by wonder.me.

East Lansing, MI, USA — Wolfgang Banzhaf
Tijuana, Baja California, Mexico — Leonardo Trujillo
Hagenberg, Austria — Stephan Winkler
Ann Arbor, MI, USA — Bill Worzel
September 2021

Contents

1 Finding Simple Solutions to Multi-Task Visual Reinforcement Learning Problems with Tangled Program Graphs 1
Caleidgh Bayer, Ryan Amaral, Robert J. Smith, Alexandru Ianta, and Malcolm I. Heywood

2 Grammar-Based Vectorial Genetic Programming for Symbolic Regression 21
Philipp Fleck, Stephan Winkler, Michael Kommenda, and Michael Affenzeller

3 Grammatical Evolution Mapping for Semantically-Constrained Genetic Programming 45
Alcides Fonseca, Paulo Santos, Guilherme Espada, and Sara Silva

4 What Can Phylogenetic Metrics Tell us About Useful Diversity in Evolutionary Algorithms? 63
Jose Guadalupe Hernandez, Alexander Lalejini, and Emily Dolson

5 An Exploration of Exploration: Measuring the Ability of Lexicase Selection to Find Obscure Pathways to Optimality 83
Jose Guadalupe Hernandez, Alexander Lalejini, and Charles Ofria

6 Feature Discovery with Deep Learning Algebra Networks 109
Michael F. Korns

7 Back to the Future—Revisiting OrdinalGP and Trustable Models After a Decade 129
Mark Kotanchek and Nathan Haut

8 Fitness First 143
W. B. Langdon

9 Designing Multiple ANNs with Evolutionary Development: Activity Dependence 165
Julian Francis Miller

10 Evolving and Analyzing Modularity with GLEAM (Genetic Learning by Extraction and Absorption of Modules) 181
Anil Kumar Saini and Lee Spector

11 Evolution of the Semiconductor Industry, and the Start of X Law 197
Andrew N. Sloss

Index 211

Contributors

Michael Affenzeller Heuristic and Evolutionary Algorithms Laboratory (HEAL), University of Applied Sciences Upper Austria, Hagenberg, Austria;
Institute for Formal Models and Verification, Johannes Kepler University, Linz, Austria

Ryan Amaral Faculty of Computer Science, Dalhousie University, Halifax, NS, Canada

Caleidgh Bayer Faculty of Computer Science, Dalhousie University, Halifax, NS, Canada

Emily Dolson BEACON Center for the Study of Evolution in Action and Department of Computer Science and Ecology, Evolutionary Biology, and Behavior Program, Michigan State University, East Lansing, MI, USA

Guilherme Espada LASIGE, Departamento de Informática da Faculdade de Ciências da Universidade de Lisboa, Lisbon, Portugal

Philipp Fleck Heuristic and Evolutionary Algorithms Laboratory (HEAL), University of Applied Sciences Upper Austria, Hagenberg, Austria;
Institute for Formal Models and Verification, Johannes Kepler University, Linz, Austria

Alcides Fonseca LASIGE, Departamento de Informática da Faculdade de Ciências da Universidade de Lisboa, Lisbon, Portugal

Nathan Haut Michigan State University, Lansing, MI, USA

Jose Guadalupe Hernandez BEACON Center for the Study of Evolution in Action and Department of Computer Science and Ecology, Evolutionary Biology, and Behavior Program, Michigan State University, East Lansing, MI, USA

Malcolm I. Heywood Faculty of Computer Science, Dalhousie University, Halifax, NS, Canada

Alexandru Ianta Faculty of Computer Science, Dalhousie University, Halifax, NS, Canada

Michael Kommenda Heuristic and Evolutionary Algorithms Laboratory (HEAL), University of Applied Sciences Upper Austria, Hagenberg, Austria;
Josef Ressel Center for Symbolic Regression, University of Applied Sciences Upper Austria, Hagenberg, Austria

Michael F. Korns Korns Associates, San Juan, PR, USA

Mark Kotanchek Evolved Analytics LLC, Rancho Santa Fe, CA, USA

Alexander Lalejini Department of Ecology and Evolutionary Biology, University of Michigan, Ann Arbor, MI, USA;
Michigan State University, East Lansing, MI, USA

W. B. Langdon Department of Computer Science, University College London, London, UK

Julian Francis Miller University of York, York, UK

Charles Ofria Michigan State University, East Lansing, MI, USA

Anil Kumar Saini University of Massachusetts Amherst, Amherst, MA, USA

Paulo Santos LASIGE, Departamento de Informática da Faculdade de Ciências da Universidade de Lisboa, Lisbon, Portugal

Sara Silva LASIGE, Departamento de Informática da Faculdade de Ciências da Universidade de Lisboa, Lisbon, Portugal

Andrew N. Sloss Arm Ltd., Washington, USA

Robert J. Smith Faculty of Computer Science, Dalhousie University, Halifax, NS, Canada

Lee Spector Amherst College, University of Massachusetts Amherst, Amherst, MA, USA

Stephan Winkler Heuristic and Evolutionary Algorithms Laboratory (HEAL), University of Applied Sciences Upper Austria, Hagenberg, Austria;
Institute for Formal Models and Verification, Johannes Kepler University, Linz, Austria

Chapter 1
Finding Simple Solutions to Multi-Task Visual Reinforcement Learning Problems with Tangled Program Graphs

Caleidgh Bayer, Ryan Amaral, Robert J. Smith, Alexandru Ianta, and Malcolm I. Heywood

Abstract Tangled Program Graphs (TPG) represents a genetic programming framework in which emergent modularity incrementally composes programs into teams of programs into graphs of teams of programs. To date, the framework has been demonstrated on reinforcement learning tasks with stochastic partially observable state spaces or time series prediction. However, evolving solutions to reinforcement tasks often requires agents to demonstrate/ juggle multiple properties simultaneously. Hence, we are interesting in maintaining a population of diverse agents. Specifically, agent performance on a reinforcement learning task controls how much of the task they are exposed to. Premature convergence might therefore preclude solving aspects of a task that the agent only later encounters. Moreover, 'pointless complexity' may also result in which graphs largely consist of hitchhikers. In this research we benchmark the utilization of *rampant mutation* (multiple mutations applied simultaneously for offspring creation) and *action programs* (multiple actions per state). Several parameterizations are also introduced that potentially penalize the introduction of hitchhikers. Benchmarking over five VizDoom tasks demonstrates that rampant mutation reduces the likelihood of encountering pathologically bad offspring while action programs appears to improve performance in four out of five tasks. Finally, use of TPG parameterizations that actively limit the complexity of solutions appears to result in very efficient low dimensional solutions that generalize best across all combinations of 3, 4 and 5 VizDoom tasks.

C. Bayer · R. Amaral · R. J. Smith · A. Ianta · M. I. Heywood (✉)
Faculty of Computer Science, Dalhousie University, Halifax, NS, Canada
e-mail: mheywood@dal.ca

C. Bayer
e-mail: caleidgh.bayer@dal.ca

R. Amaral
e-mail: ryan.amaral@dal.ca

R. J. Smith
e-mail: robert.smith@dal.ca

A. Ianta
e-mail: aianta@dal.ca

W. Banzhaf et al. (eds.), *Genetic Programming Theory and Practice XVIII*,
Genetic and Evolutionary Computation,
https://doi.org/10.1007/978-981-16-8113-4_1

1.1 Introduction

Developing policies for high-dimensional partially observable reinforcement learning tasks most often takes the form of a deep learning framework. The resulting architectures have been able to demonstrate a wide range of impressive solutions to robotics [23], gaming [10] and control problems [8]. However, solutions based on deep learning always require hardware acceleration (even post training) and the specifics of the deep learning architecture have to be designed by hand. In contrast, the tangled program graph framework represents a process for open-ended emergent modularity in which programs are rewarded for decomposing the task [11, 12, 14]. Prior research has demonstrated competitive results for TPG applied to a wide range of reinforcement learning tasks (e.g. Atari video games [11, 12], Dota 2, ViZDoom navigation [15, 20–22], and multi-task learning [13, 15]). Moreover, solutions are computationally very efficient, in part because programs have to explicitly learn what to index from the state space. That said, TPG is also limited to scalar actions and can potentially evolve to very large graphs that mostly consist of hitchhikers [9].

In this work, we investigate the impact of three factors that potentially play into diversity maintenance within the TPG framework for visual reinforcement learning while also making credit assignment less opaque: 1) multiple mutations per variation step (rampant mutation), 2) multiple actions per state (action programs) and 3) complexity limiting TPG parameterizations. Specifically, the modular structure of TPG limits the scope of mutation to single learners, thus rampant mutation affects multiple learners in multiple ways (i.e. increasing diversity). Action programs enables an agent to suggest multiple actions per state (i.e. more can potentially be done with less). Complexity limiting parameterizations might also enable us to do more with less, however, the impact of each optimization on each other needs to be established.

ViZDoom is used to provide a source of five different tasks that the visual reinforcement learning agents have to solve simultaneously (Sect. 1.4). The task domain is high-dimensional and partially observable. TPG therefore has to support both spatial and temporal representations. The canonical TPG framework assumed for this purpose is summarized in Sect. 1.2. Rampant mutation and action programs extend this framework and are described in Sect. 1.3, while the methodology adopted to encourage simple solutions through TPG parameterization appears in Sect. 1.5. An empirical evaluation follows in Sect. 1.6 with performance considered from the perspective of training fitness over the five tasks, post training multi-task evaluation over all 5 tasks, and complexity of the resulting solutions. A discussion concludes the paper with recommendations for future work (Sect. 1.7)

1.2 Tangled Program Graphs

TPG is based on a tuple $\langle T, L \rangle$ defining team and learner populations respectively [11, 12]. At initialization, each team identifies team compliment through references to a subset of learners from the learner population. Learners are rewarded for identifying the (input) context under which to apply a discrete scalar action. However, as evolution progresses variation operators enable learners to reference individuals from the team population. Such a process provides for the open ended decomposition of a task across the policies as represented by different teams. In the following, we summarize the relationship between members of the Learner (Sect. 1.2.1) and Team (Sect. 1.2.2) populations which ultimately results in the representation of solutions as a 'graph-of-teams-of-programs' (Sect. 1.2.3). Section 1.2 therefore establishes what we take to be 'canonical TPG' before we introduce two algorithmic speedups in Sect. 1.3.

1.2.1 Learners

A learner, $L(i)$, is defined in terms of a program, *prog*, and terminal action, a where $a \in A$ is the set of discrete terminal atomic actions specific to the task environment, or $L(i) = \langle p(i), a(i) \rangle$. A program, p, only produces a single output, whether that be the root node of tree structured GP [17] or register $R[0]$ in the case of linear GP [2]. Actions are merely a scalar corresponding to a terminal action (these will later evolve to also encompass pointers to other teams Sect. 1.2.2). The purpose of a program is to define context for the corresponding action. The same program can appear in different learners if it is partnered with a different action. A learner on its own does not define anything useful. Learners only appear in the Learner population, L.

1.2.2 Teams

An independent team population, T, conducts a search for good combinations of learners using a variable length representation. The following constraints are enforced: (1) each team, $tm(j)$ must consist of a unique combination of learners, $L(i) \in L$; (2) the same learner, $L(i)$, may appear in multiple teams; (3) there cannot be less than two learners in the same team; (4) there must be at least two different actions represented by the complement of learners within the same team.

In order to establish the output of team $tm(j)$, all programs associated with learners within this team are evaluated on the current input state (provided by the task environment), $\mathbf{s}_t$, or $\forall i \in tm(j) : y_i = prog_i(\mathbf{s}_t)$. The program with maximum output on $\mathbf{s}_t$ is identified by $i^* = \arg\max_i(y_i)$. Such a program wins the right to suggest its corresponding atomic action, or a_t. Under reinforcement learning tasks, the agent's action is forwarded to the task, typically resulting in a change to the task environment. Moreover, there is also a scalar reward received at the next time step, r_{t+1}. Such a reward reflects the relative significance of applying action a_t in state $\mathbf{s}_t$. Such rewards capture underlying properties of the task, such as a robot not colliding with a wall, or the robot's battery not being exhausted. Thus, the overall interaction between TPG and environment takes the form of a sequence of interactions:

$$\mathbf{s}_0, a_0, r_1, \mathbf{s}_1, a_1, r_2, \ldots, \mathbf{s}_n, a_n, r_T \quad (1.1)$$

where each TPG–environment interaction is a tuple $\langle \mathbf{s}_t, a_t, r_{t+1} \rangle$ and r_T is the terminal reward received at the task's end condition.[1] Such end conditions might reflect a failure state (point at which a game agent loses a game) or a positive outcome (winning against an opponent, solving a maze) or reflect a computational limit (maximum number of interactions). The goal of TPG is to maximize the average accumulation of rewards as sampled during the interaction defined by the sequence of Eq. 1.1. This means that only *after* encountering r_T is TPG provided with feedback. This implies that TPG is closest to Policy based Monte Carlo formulations of reinforcement learning [24], as per the majority of genetic programming applied to reinforcement learning tasks [18].

Variation operators assume that the population of TPG agents are first evaluated, ranked, and the worst *Gap* teams have been removed (a breeding cycle). Any learners that are not associated with a team, are also deleted. The remaining pool (of teams) represent potential parents, of which *Gap* are selected and cloned. Only the cloned teams are modified through crossover and mutation. Crossover selects parents pairwise from the pool of surviving teams with uniform probability. The learners common to both appear in both the offspring. Learners unique to each parent are selected to appear in an offspring with probability P_{cpy}. Let the result of this process be the set of Gap offspring teams, L'.

Mutation takes the form of stochastically adding (P_a) or deleting (P_d) learners from the offspring pool, L' (subject to the above constraints). Variation is performed relative to learners indexed by teams from L' with probability P_m. Again, should a learner be selected for variation, it is first cloned. This means that only the offspring team inherits the modified learner, $L'(i)$, not any of the $T - Gap$ grandfathered teams that happened to use the same learner as a parent. Learner variation operators include: instruction delete (P_{del}), add (P_{add}), swap (P_{swp}) and choose a new terminal action (P_{mn}).

1.2.3 Graphs

Section 1.2.1 defined a learner as the smallest 'module' whereas Sect. 1.2.2 provided a mechanism for organizing learners into teams without prior parameterization for how many learners should appear in a team.[2] Different teams might excel at defining policy for different subsets of the state-action sequence. Typically, it is assumed that crossover will provide a sufficient mechanism for recombining the properties from different teams. The underlying premise to this is that the learners when merged using crossover continue to identify unique conditions under which to out-bid other

[1] Implies that the interaction represents the special case of an episodic task [24].

[2] Although a minimum of two learners (with different actions) is necessary to avoid defining a degenerate team Sect. 1.2.2.

learners. Unfortunately, there is no guarantee that this will be the case. TPG may avoid this condition by enabling a learner to instead reference a different team, thus devolving control to the referenced team under state $\mathbf{s}_t$.

The key to this process is to provide two *types* of learner action mutation. At initialization all learners are initialized from a discrete set of atomic actions, $a(i) \in A$, specific to the task (e.g. an enumeration of all joystick directions). Thereafter, an action mutation consists of the sequence of tests summarized by Algorithm 1.1. Step 1 determines whether to apply any form of mutation. When true either an action from the set of atomic actions, A, is chosen (Step 5) or a pointer to another team, T, is established (Step 6). The significance of Step 4 is that it potentially forces a change in the *type* of action.

Algorithm 1.1 Mutating the action type. The 'Choose' function selects an action of the corresponding type with uniform probability.

1: **if** $rand > P_{mn}$ **then**
2: no mutation
3: **else**
4: **if** $rand > P_{action}$ **then**
5: $a(i) \leftarrow$ Choose(A)
6: **else** $a(i) \leftarrow$ Choose(T)
7: **end if**
8: **end if**

Two types of team are now recognized. Those that receive at least one reference from another team and those that do not; the latter define the set of 'root teams' T_{root}. At initialization $T_{root} = T$. Evaluation may only commence from a root team. Team evaluation is unchanged relative to that established in Sect. 1.2.2. Should the winning learner's action be an atomic action, a_t, then the action is returned to the environment, resulting the next reward, r_{t+1}, and, assuming that $r_{t+1} \neq r_T$, the next state $\mathbf{s}_{t+1}$. Otherwise, the action is a pointer to another team and the process of determining the winning learner repeats at the identified team.

The set of eligible parents is also limited to the set of root teams, or $tm \in T_{root}$. Thus, variation operators (Sect. 1.2.2) are *only* applied to root teams with the ratio of root to non-root teams floating. Moreover, the non-root teams essentially behave as if they have been archived, unless at some point the variation operators remove all incoming references. Note, however, that this has no impact on the pool of actions that action mutation may select from, only which teams can be parents.

Naturally, it is also possible for loops to appear in the path of evaluation, i.e. the halting problem. TPG avoids this issue, by marking teams visited during the evaluation of a root team. Should a learner identify a previously visited team, then the learner with runner up bid is (recursively) selected. By enforcing the constraint that all teams have to have a minimum of one terminal action, TPG guarantees that loops cannot result.[3] Further details of the TPG algorithm appear in tutorial form in [12, 14].

1.2.4 Memory

The partially observable aspects of the ViZDoom task imply that support for memory is beneficial [22], even with respect to single ViZDoom source tasks [15]. For the purposes of this study, we will adopt the probabilistic indexed memory formulation previously benchmarked under ViZDoom and Dota 2 reinforcement learning environments [15, 21, 22]. In summary, only one instance of indexed memory is retained. This implies that a TPG agent inherits the state of indexed memory left by the previous agent. Indexed memory therefore represents a global internal model of state that is never reset. Registers, R, specific to a learner (Sect. 1.2.1) are considered to capture the internal state of each learner. With this in mind, the instruction set is augmented with a write (`write(R)`) and read (`R[i] = read(k)`) operation. Write operations are probabilistic, distributing the content of a learner's registers across L columns of indexed memory. The probability of performing a write is such that locations towards column 1 and L are less likely to be written to (or long term memory). Conversely, locations near $\frac{L}{2}$ are most likely to be written to (or short term memory). Read operations specify a target register, $R[i]$, and an 'address' (k) to indexed memory, i.e. $0 < k \leq L \times \texttt{MaxReg}$. Further details of the probabilistic indexed memory model can be found in earlier work [15, 21, 22].

1.3 Mechanisms for Accelerating TPG Evolution

Two mechanisms are investigated for accelerating the operation of TPG: Rampant Mutation (Sect. 1.3.1) and Multi-actions (Sect. 1.3.2).

1.3.1 Rampant Mutation

Rampant mutation represents a burst of mutation that occurs on certain generations as a means of increasing the potential for greater diversity. To this end, let it $M(\cdot)$ denote the rampant mutation operation that accepts two parameters (a, b) to configure the mutation behaviour, where a is an integer generation and b is a mutation multiplier. Thus, $M(a, b)$ is interpreted as "Every a-th generation, perform mutation b times

[3] An arc marking scheme has since been proposed [9], however, for the purpose of this work the original team formulation was assumed.

instead of once". Assuming a parameterization of $M(1, 5)$ would therefore result in $\times 5$ the base level of mutation at every generation.

Relative to prior work, Cobb defined a 'hypermutation' operator as the application of different levels of mutation during evolution in proportion to fitness, i.e. decreases in fitness trigger hypermutation [4]. Conversely, Grefenstette replaced a percentage of the population at each generation with randomly generated individuals or random immigrants [7]. Ghosh et al. gave more reproductive rights to agents in a certain age range [6]. The prior works therefore established that 'diversity maintenance' (care of rates of variation) could be useful under dynamic environments, albeit with a much simpler genotype and a fixed length representation. In this work, we are interested in knowing whether the combination of higher-levels of variation in programs (care of rampant mutation) adversely or positively impact on the ability of TPG to construct useful graphs. In applying the variation throughout evolution we recognize that waiting until fitness stagnates before attempting to introduce diversity may be too late to correct for a loss in diversity [3]. We also note that, unlike the earlier studies, TPG is explicitly modular. Hence mutation when applied is specific to an affected learner (aka module), where a candidate solution must consist of a minimum of two learners (but in practice might comprise from thousands).

Multiple 'rampant' forms of mutation occur in biological organisms. For example, the immune system deploys targeted mutations to the immunoglobulin genes in a process referred to as somatic hypermutation [25]. The underlying objective is to respond to threats experienced by an individual, i.e. an adaptive mechanism for programming/targeting mutation during the lifetime of the individual. A second example is the case of stress-induced mutations in microbes. These are interesting because they occur when a microbe is poorly adapted to its environment [1].

1.3.2 Multi-actions

A canonical TPG learner, $L(i)$, is defined by the tuple $\langle p(i), a(i) \rangle$ or program, p, (bidding behaviour) and scalar action, a, (Sect. 1.2.1), thus limiting the tasks to which TPG could be applied to as those with discrete actions alone. In order to provide support for multiple real-valued actions *per learner*, we introduce a new representation $\langle p_B(i), p_A(i), a(i) \rangle$ in which $p_B(i)$ is the bid program (operation unchanged), $p_A(i)$ is the atomic action program, and $a(i)$ is the pointer to a team. Naturally, at any point in time, the learner has enabled *either* $p_A(i)$ or $a(i)$, never both.

The purpose of the bidding program, $p_B(i)$, is unchanged relative to that of canonical TPG (Sect. 1.2.2). If the learner's action is a Team reference ($a(i)$ enabled), then graph traversal follows the same process as outlined above (Sect. 1.2.3). Conversely, if the learner's action program is enabled, then $p_A(i)$ is executed (relative to the same environmental state, $\mathbf{s}_t$). In order to efficiently support multiple actions *per state* we assume a linear GP representation [2]. Post execution, the action programs registers represent a vector of actions. Thus, as long as the number of registers per action program `MaxActReg` is at least as many as required by the task, then the numerical value in register `R[i]` is the value for atomic action i of the task under state $\mathbf{s}_t$.

As each learner consists of two programs, credit assignment is potentially more complex. A hierarchical process is therefore assumed in which action program mutation is conditional on the corresponding bidding program having been first mutated.

Relative to previous work, an example of real-valued TPG has been previously proposed by Kelly et al. [13] and benchmarked on *single* output time series prediction tasks. Conversely, in this work multiple actions are always necessary. Note that for the ViZDoom task discrete actions will be assumed in which case all action program registers greater than zero imply an action is enabled, otherwise an action is not enabled at that state. A future benchmarking study will consider the case of real-valued multi-actions per state.

1.4 ViZDoom Subtask Selection and Performance Evaluation

The ViZDoom game engine [16] represents a 3D environment through first person perspective, and therefore a challenging visual reinforcement learning task on account of: (1) the high-dimensionality and partial observability of state, $\mathbf{s}(t)$, (2) the multitude of different objects/opponents all of which can appear at multiple aspect ratios and/or view angle, and (3) the environment is stochastic, with the spawn state of the agent and opponents changing. Moreover, the environment comes with a set of eight default subtasks [16] that can form the basis for training curricula [20].

Past experience with ViZDoom subtasks has demonstrated that a lack of diversity can limit the development of agents within the context of training on a single subtask. With this in mind, we will therefore adopt the approach of evolving agents to play five subtasks simultaneously by randomly choosing a subtask (without replacement) and evaluating performance on the three most recently encountered subtasks [20].

In short, we have a 'bag' of subtasks, $B = 5$, each of which can be chosen with equal probability (without replacement). On choosing subtask S, then the *same* subtask is evaluated $\tau = 5$ times per agent.[4] Resulting in the average performance of agent i on a subtask S:

$$f(i, S) = \frac{1}{\tau} \sum_{k=0}^{\tau-1} gs_k(i, S) \tag{1.2}$$

where $gs_k(i, S)$ is the game score returned by the ViZDoom game engine on agent i encountering a terminal reward condition, r_T, (i.e. episodic reinforcement Eq. (1.1)) under evaluation k for subtask S.

[4] Stochastic nature of each subtask requires that agents are evaluated over multiple initializations.

Each subtask has a completely different performance scale for their subtask (Sect. 1.5.1). Thus, the subtask specific score is rescaled relative to the best agent score on that subtask:

$$F(i, S) = \frac{f(i, S)}{\max_{j \in T_{root}} f(j, S)} \quad (1.3)$$

where T_{root} denotes the set of eligible agents, i.e. the set of root teams (Sect. 1.2.3).
Performance of agent i is now expressed in terms of the performance across the current plus $R = 2$ previous subtasks encountered:

$$F(i) = \sum_{j=0}^{R} F(i, S - j) \quad (1.4)$$

Such a process exposes agents to switches in subtask, where such switches have previously demonstrated to be effective at promoting development of more general agent behaviours [19]. Thus, by exposing agents to multiple subtasks, we make use of the population to act as a repository of multiple agent behaviours that TPG ultimately is rewarded for 'stitching together' to compose agents with a multitude of skills [20, 22].

1.5 Empirical Methodology

1.5.1 Task Domains

TPG agents will be evolved against total of 5 subtasks from the ViZDoom game engine: Basic, Health Gathering, Defend the Centre, Defend the Line, and Take Cover.[5] The objectives and eligible actions per task varies, but (input) state at each time step, $\mathbf{s}(t)$, always takes the form of a 160×120 resolution RGB frame from the game engine. Each colour channel takes the form of an 8-bit integer that is then concatenated into a single 24-bit integer [20].

Basic is designed to develop basic aiming skills. Each episode begins by spawning the agent in the centre of the long side of a rectangular room. A monster is initialized on the opposing wall at a random location. The agent can only turn left or right and shoot. Rewards are -1 per elapsed time step, -5 for each shot that misses the monster and +101 for a shot that hits the monster. The episode terminates when the monster is shot or 300 time steps (frames) elapse.

Defend the centre generalizes the skill from 'Basic' by spawning the agent in the middle of a rectangular room. The agent has a limited amount of ammunition and opposes 5 monsters simultaneously. The monsters also re-spawn, thus the objective

[5] https://github.com/mwydmuch/ViZDoom/tree/master/scenarios.

is to live for as long as possible. The agent is only allowed to turn left/right and shoot. The reward is +1 for each monster killed, whereas the episode ends when the agent is killed.

Defend the line initializes the agent in a rectangular room as under 'Basic', but this time with three monsters on the opposite wall. In this scenario, monsters initially die after one 'on-target' shot, but on re-spawning (in a random location on the wall opposing the agent) have more 'strength', hence need to be hit more times before dying. The rewards and episode end conditions are the same as 'Defend the centre'.

Health gathering also assumes a rectangular room, however, there are no monsters. Instead, the floor is 'acidic' thus decreases the health of the agent. In order to survive the agent has to pick up health packs which add a limited amount of health back to the agent. The health packs randomly spawn over the course of the episode. The agent can move forward and turn left or right. An agent receives a reward of +1 for each frame for which it survives. The episode terminates if the agent dies or the agent successfully survives for 2,100 frames.

Take cover assumes the same starting condition as Basic and Defend the line. Monsters are again randomly spawned on the opposing wall, but this time can launch fire-balls at the agent. The agent can only move left or right. Agents receive a reward of +1 for each time step they survive. The longer the agent survives the more monsters are spawned, thus at some point the agent faces too many monsters to survive any longer.

In short, the tasks are related, but reward different behaviours. TPG agents are therefore initially rewarded for discovering solutions to individual sub-tasks. However, the performance function is formulated to reward behaviour over a sequence of the last three sub-tasks encountered (Eq. (1.4)). This encourages the TPG to develop behaviours that generalize over any combination of sub-task. Note that there is no 'special flag' that uniquely identifies what each sub-task is, only the visual information from the game engine.

1.5.2 Parameters

A total of five TPG configurations will be considered, as summarized in Table 1.1. Previous research benchmarked TPG without either rampant mutation or action programs [20]. **NRAP** represents the case of no rampant mutation, but action programs are enabled. This is the only case that does not employ rampant mutation.[6] **RAL** enables rampant mutation, but no action programs (zero for 'Action Registers'). **RAP** enables both rampant mutation and action programs. **RAPF** assumes RAP, but delays TPG from indexing other teams until generation 3,000. The motivation for this is to force single teams to first get 'strong' before graphs are constructed that potentially stitch multiple strong teams together. Finally, **RAPS** also assumes RAP, but limits the size of an initial team to 2 programs as well as limiting the maximum

[6] Reflected in the parameterization of the 'Rampant Magnitude' row in Table 1.1.

Table 1.1 Configuring TPG with and without rampant mutation and multi-action programs

Parameter	NRAP	RAL	RAP	RAPF	RAPS
Team pop size	120				
Team gap	0.5				
Prob learner delete	0.7				
Prob learner add	0.7				
Prob mutate action	0.2				
Prob mutate team ref	0.5				
Maximum team size	12	12	12	12	**4**
Maximum program size	128				
Prob program delete	0.5				
Prob program add	0.5				
Prob program swap	1.0				
Prob program mutate	1.0				
Maximum action program size	128				
Prob action program delete	0.5				
Prob action program add	0.5				
Prob action program swap	1.0				
Prob action program mutate	1.0				
Number of action registers	7	**0**	7	7	7
Starting learners per team	12	12	12	12	**2**
Phase flip	0	0	0	**3000**	0
Episodes	5				
Generations	9,000				
Rampant magnitude	**1**	5	5	5	5
Rampant frequency	1				
Screen resolution	160 × 120 (19,200 pixels)				

size of a team to 4 programs. Thus, all the other configurations start with a 'full compliment' of 12 programs per team (variation operators can decrease this), so the RAPS configuration requires TPG to construct solutions from multiple teams, whereas the others need not. In the case of indexed memory, the same parameterization is assumed as with prior work [15, 21, 22]. Thus, indexed memory consists of $L = 100$ columns for a total of $\mathbf{m} = 100 \times \texttt{MaxReg}$ cells, where `MaxReg` is 8.

1.6 Results

1.6.1 Fitness

Fitness curves are used to express development of the TPG configurations. It takes approximately a week to perform 1,000 generations, where each generation is a complete pass through the 'bag' of source tasks (Sect. 1.4). Figure 1.1 illustrates the development of the 5 tasks over the 9,000 training generations for a typical run. In each case, the shaded region of the plot illustrates the performance spread (best to worst) of the champion agent under each task (average is the dashed line). The solid curve is the average performance of the population as a whole. Note also that the scales for each task are very different, reflecting the different reward schemes (Sect. 1.5.1).

We note that progress was made most consistently on the Basic and Health Gathering tasks (max. of 100 and 2,100 respectively). It is apparent that only RAL was unable to identify a champion agent that could not reach the maximum score under Health Gathering (Fig. 1.1i). Indeed, the performance of RAL (green curves) consistently returned the best performance of all 5 configurations under Defend the Line (Fig. 1.1g), but the worst at Defend the Center (Figure 1.1c) and Health Gathering. The configuration without rampant mutation (NRAP, red curves) was interesting because it appeared to collapse towards the end of the run. Conversely, the two most constrained TPG configurations (RAPF and RAPS; purple and orange curves) do not appear to loose anything relative to the 'unconstrained' TPG configurations, possibly implying that it is generally best to stay simple in this subset of tasks.

1.6.2 Generalization

Post training evaluation identifies the champion from training and re-assesses under each task using 50 initializations per task. From the perspective of generalization, we are interested in the ability of agents to perform multiple tasks simultaneously. With that in mind we construct all combinations of three, four and five tasks and then construct a table of the relative ranking of each TPG configuration. This enables us to identify whether a statistically significant outcome appears and having rejected the null hypothesis (all ranks are the same), apply a post hoc test for which configurations perform differently. Table 1.2 summarizes the outcome of the ranking across each task combinations.

Applying the Friedman non-parametric test determines the likelihood of the distribution of ranks being different from the average rank (3 in this case),

$$\chi_F^2 = \frac{12\,N}{k(k+1)} \left[\sum_j R_j^2 - \frac{k(k+1)^2}{4} \right] \quad (1.5)$$

where k is the number of TPG configurations (5) and N is the number of tasks (16). Specifically, under the ranks from Table 1.2 returns a $\chi_F^2 = 18.55$. This is then renormalized to provide a critical value distributed according to the F-distribution [5] using,

Fig. 1.1 Fitness curves during training. NRAP (no rampant mut., but with action programs) are the red curves. RAL (rampant mut., discrete actions) are the green curves. RAP are the blue curves (rampant mut., with action programs). RAPF and RAPS are purple and orange curves (phased TPG and small TPG initialization)

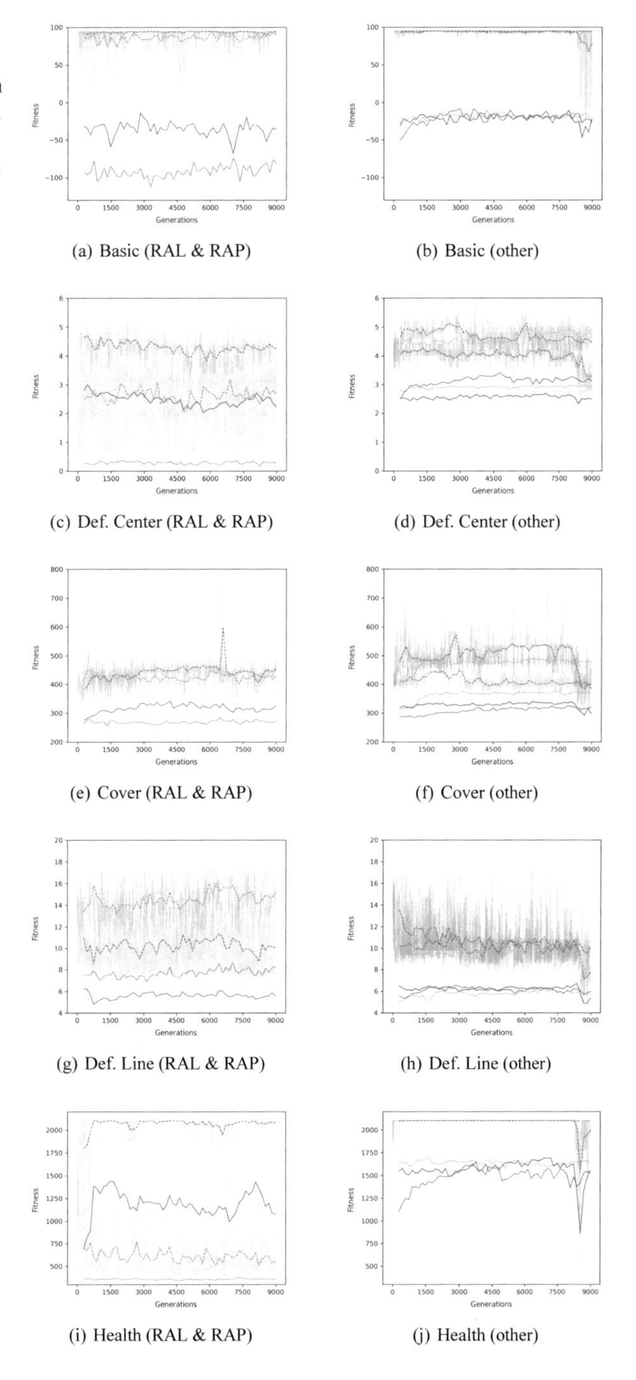

(a) Basic (RAL & RAP)

(b) Basic (other)

(c) Def. Center (RAL & RAP)

(d) Def. Center (other)

(e) Cover (RAL & RAP)

(f) Cover (other)

(g) Def. Line (RAL & RAP)

(h) Def. Line (other)

(i) Health (RAL & RAP)

(j) Health (other)

Table 1.2 Combinations of 3, 4, and 5 tasks versus TPG configuration. ViZDoom tasks are identified as 'b' for basic; 'hg' for health gathering, 'dtl' for defend-the-line, 'dtc' for defend-the-center, 'tc' for take cover

	TPG configuration				
Task	RAL	RAP	NRAP	RAPF	RAPS
b-hg-dtl	5	4	2	1	3
b-hg-dtc	5	2	3	1	4
b-hg-tc	5	3	4	2	1
b-dtl-dtc	5	4	3	1	2
b-dtl-tc	1	4	5	3	2
b-dtc-tc	5	2	4	3	1
b-dtl-dtc	5	4	2	1	3
hg-dtl-tc	5	4	3	2	1
hg-dtc-tc	5	3	4	1	2
dtl-dtc-tc	3	4	5	2	1
b-hg-dtl-dtc	5	4	2	1	3
b-hg-dtl-tc	5	4	3	2	1
b-hg-dtc-tc	5	3	4	1	2
b-dtl-dtc-tc	5	3	4	2	1
hg-dtl-dtc-tc	5	3	4	1	2
all 5 tasks	5	3	4	1	2
avg. rank (R)	4.25	3.375	3.5	1.5625	1.9375

$$F_F = \frac{(N-1)\chi_F^2}{N(k-1) - \chi_F^2} \tag{1.6}$$

returning $F_F = 6.12$ with $k - 1 = 4$ and $(k-1)(N-1) = 60$ degrees of freedom, thus $F(4, 60) = 2.53$. As $F(4, 60) < F_F$ the the null hypothesis can be rejected.

The Bonferroni-Dunn post hoc test can now be applied to establish the critical difference (CD) between pairs of TPG configurations [5]. Thus, defining $CD = q_\alpha\sqrt{(\frac{k(k+1)}{6N})}$ and looking up the critical value for 5 models as $q_{0.05} = 2.498$ then $CD = 1.398$. From this we can now deduce that RAPF and RAPS are significantly different from RAL and NRAP.

1.6.3 Complexity

In addition to tracking the dynamic properties of fitness over the generations (Sect. 1.6.1), we can also investigate the development of complexity. With that in mind, three metrics will be tracked: Learners per Team, Instructions per Learner, and Instructions per team. Figure 1.2a summarizes development through the average Learners per

Team under the five TPG configurations. Three trends are apparent, RAPS (orange) retain the least number of Learners per Team, retaining less than half the team members of any other TPG configuration. Naturally, this reflects the maximum learner per team parameterization specific to RAPS (Table 1.1). RAPF (purple) appears to represent a second level of complexity with an average of about 9 Learners per Team. NRAP, RAL and RAP represent the most complex teams with 10 learners per team on average. Moreover, NRAP (red) appear to undergo a switch in complexity (to a much lower value) around the 8,500 generation mark, where this seems to have also caused a collapse in the corresponding fitness curves (Sect. 1.6.1).

Figure 1.2b characterizes the development of the average number of instructions per learner. RAPS (orange) now represent the most complex individuals, implying that the low team complement observed under Fig. 1.2a is being compensated by larger programs.[7] Conversely, RAPF (purple) consistently adopts the lowest program complexity after a period of complexity around the first 500 generations. NRAP (red) appear very quickly converge on an instruction count just less than 70 instructions, but again at the 8,500 generation mark display a switch (to a much higher value). RAP consistently assumed solutions with 70 instructions after about 1,000 generations, whereas RAL (green) displayed the most variation in the average instructions per learner over the duration of the run.

Figure 1.2c summarizes complexity from the perspective of the instructions per team. Again RAPS represents the least complexity, in part reflecting the team size parameterization of no more than 4 learners per team. RAPF (purple) identified the next lowest instruction per team complement, where this reflects the low instruction to learner complement. RAL were again the most complex with NRAP and RAP for the most part adopting a similar value. The one exception being NRAP around the 8,500 generation mark where a collapse in complexity followed by a recovery takes place.

In short, the RAL configuration of TPG appear to represent the most complex solutions. RAL represents the case of discrete action labels where this appears to imply that more complexity is necessary to offset the scalar action constraint. NRAP represents the scenario without rampant mutation, and was the only case to demonstrate a sudden collapse in solution complement. Rampant mutation by modifying multiple components of offspring might therefore provide more opportunity for repairing negative variation. RAP demonstrated most complexity, whereas both of the 'constrained' TPG parameterizations with action programs (RAPS and RAPF) where able to control complexity without impacting on the ability to identify quality solutions.

[7] Includes introns and hitchhikers.

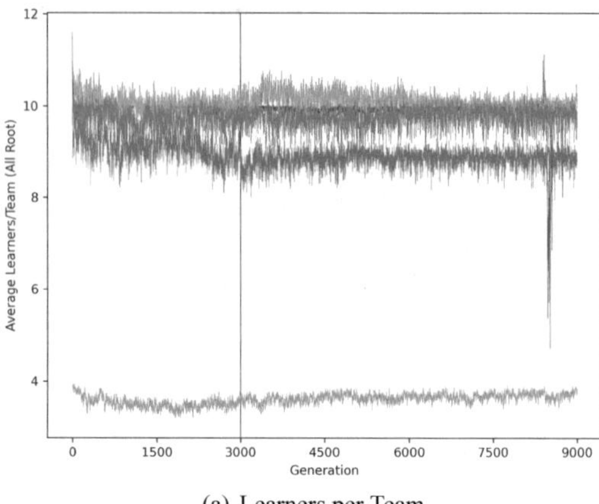

(a) Learners per Team

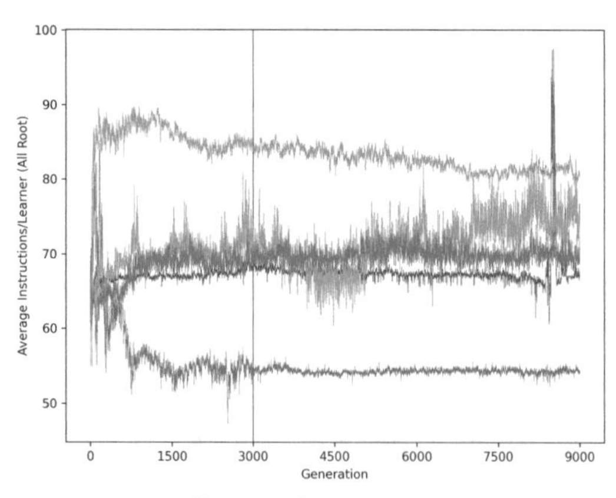

(b) Instructions per Learner

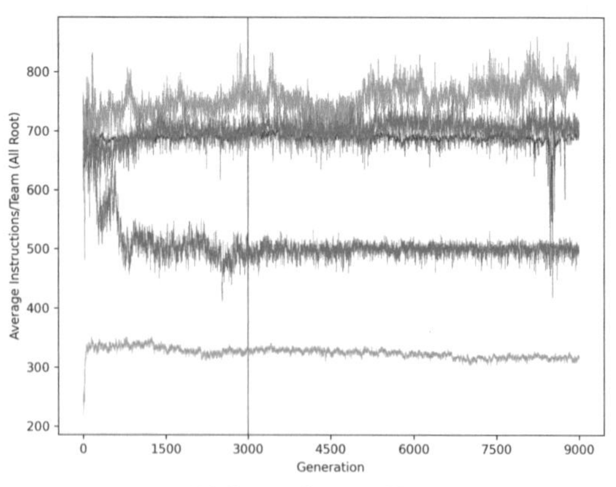

(c) Instructions per Team

Fig. 1.2 Complexity of champions. Vertical line at 3,000 generations indicates point at which TPG allowed to index other teams under RAPF. NRAP (no rampant mut., but with action programs) are the red curves. RAL (rampant mut., discrete actions) are the green curves. RAP are the blue curves (rampant mut., with action programs). RAPF and RAPS are purple and orange curves (phased TPG and small TPG initialization)

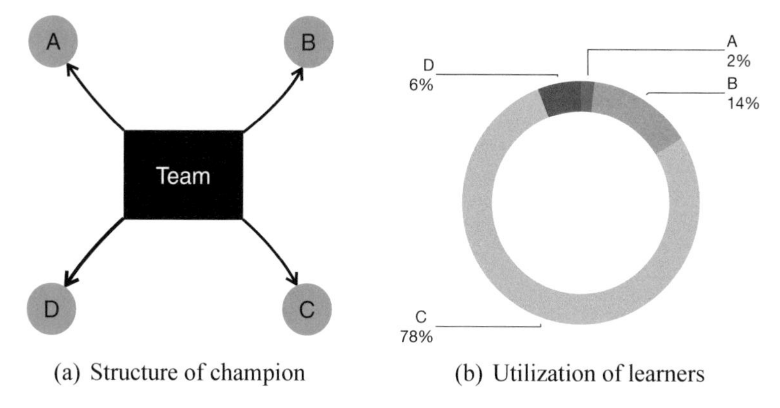

Fig. 1.3 Simplest RAPS champion solution. **a** Individual consists of four learners $\langle p_B(i), p_A(i)\rangle$. **b** Utilization of each learner across the five ViZDoom tasks

1.6.4 Details of a RAPS Solution

Section 1.6.3 established that the RAPS parameterization of TPG typically consisted of a single team with no more than four programs. Moreover, generalization performance did not appear to be adversely affected (Sect. 1.6.2). With that in mind we can also analyze the nature of a RAPS champion in more detail. Figure 1.3a summarizes the basic structure of such a solution in which $\langle A, B, C, D\rangle$ are learners each comprising of a unique bidding and action program $\langle p_B(i), p_A(i)\rangle$. Figure 1.3b summarizes the corresponding utilization of each learner across all tasks, where the standard deviation between tasks is $< 0.6\%$. Interestingly, the number of instructions in a bid program is either 78 or 79 and between 108 to 110 for action programs (including introns). What is interesting is that given the high dimension of the state space ($160 \times 120 = 19, 200$), the resulting TPG agents are only indexing 19 to 22 pixels for the action programs and 10 to 12 pixels for the bidding programs. Visually inspecting the resulting source code indicates that action programs do share similar code structures with each other (albeit with different register references and switched instruction orders), whereas bidding programs have a lot less in common.

1.7 Conclusions

Tangled program graphs has previously been applied to several visual reinforcement learning tasks (Arcade Learning Environment [11, 12, 14] and ViZDoom [14, 15, 20, 22]). In this work, various parameter and algorithmic optimizations are benchmarked on a set of ViZDoom sub-tasks. The underlying goal is to evolve agents able to play each sub-task from a single evolutionary run. A preference appears to exist for action programs with statistically significantly better generalization demonstrated when both action programs and limited TPG parameterizations are adopted. The action program takes the form of linear GP, and as such can express multiple outputs simultaneously. Previously, TPG actions were limited to a single discrete value. This also appears to imply that TPG solutions can be simpler (RAPF and RAPS configurations) without impacting on performance. Rampant mutation appears to be less sensitive to introducing pathological modifications to TPG structure, however, some form of elitism might also be beneficial in this regard. Future research will continue to develop the TPG framework, with the introduction of mutation operators that apply horizontal/vertical offsets or rotations to the pixels indexed by a program being of particular interest.

Acknowledgements We gratefully acknowledge support from the NSERC CRD and Discovery programs (Canada).

References

1. Bjedov, I., Tenaillon, O., Gerard, B., Souza, V., Denamur, E., Radman, M., Taddei, F., Matic, I.: Stress-induced mutagenesis in bacteria. Science **300**, 1404–1409 (2003)
2. Brameier, M., Banzhaf, W.: Linear Genetic Programming. Springer (2007)
3. Branke, J.: Evolutionary approaches to dynamic environments—a survey. In: GECCO Workshop on Dynamic Optimization Problems, pp. 134–137 (1999)
4. Cobb, H.G.: An investigation into the use of hypermutation as an adaptive operating in genetic algorithms having continuous, time-dependent non-stationary environments. Technical Report TR AIC-90-001, Naval research Laboratory (1990)
5. Demsar, J.: Statistical comparisons of classifiers over multiple data sets. J. Mach. Learn. Res. **7**, 1–30 (2006)
6. Ghosh, A., Tstutsui, S., Tanaka, H.: Function optimization in non-stationary environment using steady state genetic algorithms with aging of individuals. In: IEEE Congress on Evolutionary Computation, pp. 666–671 (1998)
7. Grefenstette, J.J.: Genetic algorithms for changing environments. In: PPSN, pp. 137–144 (1992)
8. Hwangbo, J., Lee, J., Dosovitskiy, A., Bellicoso, D., Tsounis, V., Koltun, V., Hutter, M.: Learning agile and dynamic motor skills for legged robots. CoRR (2019). arXiv:abs/1901.08652
9. Ianta, A., Amaral, R., Bayer, C., Smith, R.J., Heywood, M.I.: On the impact of tangled program graph marking schemes under the atari reinforcement learning benchmark. In: Proceedings of the ACM Genetic and Evolutionary Computation Conference, p. to appear (2021)
10. Jaderberg, M., Czarnecki, W.M., Dunning, I., Marris, L., Lever, G., Castañeda, A.G., Beattie, C., Rabinowitz, N.C., Morcos, A.S., Ruderman, A., Sonnerat, N., Green, T., Deason, L., Leibo, J.Z., Silver, D., Hassabis, D., Kavukcuoglu, K., Graepel, T.: Human-level performance in 3D multiplayer games with population-based reinforcement learning. Science **364**, 859–865 (2019)
11. Kelly, S., Heywood, M.I.: Emergent tangled graph representations for atari game playing agents. In: European Conference on Genetic Programming, LNCS, vol. 10196, pp. 64–79 (2017)
12. Kelly, S., Heywood, M.I.: Emergent solutions to high-dimensional multitask reinforcement learning. Evol. Comput. **26**(3), 347–380 (2018)
13. Kelly, S., Newsted, J., Banzhaf, W., Gondro, C.: A modular memory framework for time series prediction. In: Proceedings of the ACM Genetic and Evolutionary Computation Conference, pp. 949–957 (2020)
14. Kelly, S., Smith, R.J., Heywood, M.I.: Emergent policy discovery for visual reinforcement learning through tangled program graphs: a tutorial. In: Banzhaf, W., Spector, L., Sheneman L

(eds.) Genetic Programming Theory and Practice XVI, Genetic and Evolutionary Computation, pp. 37–57 (2018)
15. Kelly, S., Smith, R.J., Heywood, M.I., Banzhaf, W.: Emergent tangled program graphs in partially observable recursive forecasting and ViZDoom navigation tasks. ACM Trans. Evol. Learn. Optim. **1** (2021)
16. Kempka, M., Wydmuch, M., Runc, G., Toczek, J., Jaskowski, W.: ViZDoom: A Doom-based AI research platform for visual reinforcement learning. In: IEEE Conference on Computational Intelligence and Games, pp. 1–8 (2016)
17. Koza, J.R.: Genetic Programming—On the Programming of Computers by Means of Natural Selection. MIT Press, Complex Adaptive Systems (1993)
18. Moriarty, D.E., Schultz, A.C., Grefenstette, J.J.: Evolutionary algorithms for reinforcement learning. J. Artif. Intell. Res. **11**, 199–229 (1999)
19. Parter, M., Kashtan, N., Alon, U.: Facilitated variation: how evolution learns from past environments to generalize to new environments. PLOS Comput. Biol. **4**(11), 1–15 (2008)
20. Smith, R.J., Heywood, M.I.: Scaling tangled program graphs to visual reinforcement learning in ViZDoom. In: European Conference on Genetic Programming, Lecture LNCS, vol. 10781, pp. 135–150 (2018)
21. Smith, R.J., Heywood, M.I.: Evolving Dota 2 shadow fiend bots using genetic programming with external memory. In: Proceedings of the ACM Genetic and Evolutionary Computation Conference, pp. 179–187 (2019)
22. Smith, R.J., Heywood, M.I.: A model of external memory for navigation in partially observable visual reinforcement learning tasks. In: European Conference on Genetic Programming, LNCS, vol. 11451, pp. 162–177 (2019)
23. Sünderhauf, N., Brock, O., Scheirer, W.J., Hadsell, R., Fox, D., Leitner, J., Upcroft, B., Abbeel, P., Burgard, W., Milford, M., Corke, P.: The limits and potentials of deep learning for robotics. Int. J. Robot. Res. **37**(4–5), 405–420 (2018)
24. Sutton, R.S., Barto, A.G.: Reinforcement Learning: An Introduction. MIT (2018)
25. Teng, G., Popavasiliou, F.N.: Immunoglobulin somatic hypermutation. Annu. Rev. Genet. **41**, 107–120 (2007)

Chapter 2
Grammar-Based Vectorial Genetic Programming for Symbolic Regression

Philipp Fleck, Stephan Winkler, Michael Kommenda, and Michael Affenzeller

Abstract Vectorial Genetic Programming (GP) is a young branch of GP, where the training data for symbolic models not only include regular, scalar variables, but also allow vector variables. Also, the model's abilities are extended to allow operations on vectors, where most vector operations are simply performed component-wise. Additionally, new aggregation functions are introduced that reduce vectors into scalars, allowing the model to extract information from vectors by itself, thus eliminating the need of prior feature engineering that is otherwise necessary for traditional GP to utilize vector data. And due to the white-box nature of symbolic models, the operations on vectors can be as easily interpreted as regular operations on scalars. In this paper, we extend the ideas of vectorial GP of previous authors, and propose a grammar-based approach for vectorial GP that can deal with various challenges noted. To evaluate grammar-based vectorial GP, we have designed new benchmark functions that contain both scalar and vector variables, and show that traditional GP falls short very quickly for certain scenarios. Grammar-based vectorial GP, however, is able to solve all presented benchmarks.

P. Fleck (✉) · S. Winkler · M. Kommenda · M. Affenzeller
Heuristic and Evolutionary Algorithms Laboratory (HEAL), University of Applied Sciences Upper Austria, Softwarepark 11, 4232 Hagenberg, Austria
e-mail: philipp.fleck@fh-hagenberg.at

P. Fleck · S. Winkler · M. Affenzeller
Institute for Formal Models and Verification, Johannes Kepler University, Altenberger Straße 69, 4040 Linz, Austria

M. Kommenda
Josef Ressel Center for Symbolic Regression, University of Applied Sciences Upper Austria, Softwarepark 11, 4232 Hagenberg, Austria

W. Banzhaf et al. (eds.), *Genetic Programming Theory and Practice XVIII*,
Genetic and Evolutionary Computation,
https://doi.org/10.1007/978-981-16-8113-4_2

2.1 Introduction

Symbolic Regression (SR) models, created via Genetic Programming (GP), are typically trained with input variables that represent single, real-valued values. In many applications, however, higher dimensional data is available, which requires prior feature engineering to extract relevant information that is then used as regular, real-valued features. But even when feature engineering is done with the help of domain experts, some information is always lost and important features forgotten or overlooked. Instead, a symbolic model should perform the extraction itself, which has not only the benefit of not requiring domain knowledge, but also can shed light on relations that were previously unknown and which can be easily read of a symbolic model due to its white-box nature.

As a demonstration example, we introduce a small, artificial production plant that produces items, shown in Fig. 2.1. For each produced item, we know certain input parameters such as material properties, and monitor the production process via sensors that yield continuous measurements. After the production of an item it's quality is measured. Now, symbolic regression should be used to predict the quality of the product, based on the available input data: regular, scalar values, such as input material parameters, and the continuous time series of the processes while the item passed through the production plant, represented as vector values. Table 2.1 shows how the raw input data looks like for this fictional production plant. In this case, there are two scalar variables available, *MatA* and *MatB*, while the process variables *Temp* and *Press*, representing the temperature and pressure process during the production, are vector variables. The target variable *Quality* is a regular, scalar variable.

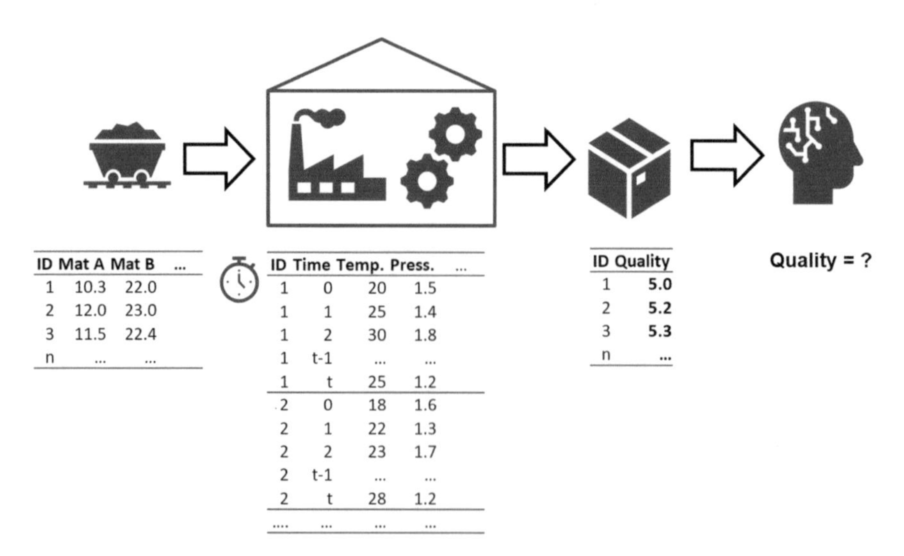

Fig. 2.1 An example of a small factory, where the product quality should be predicted based on known material properties and the measured process data during production

Table 2.1 Example data of an artificial production plant containing scalar and vector input variables, and a scalar target variable

ID	Mat A	Mat B	Temp.	Press.	Quality
1	10.3	22.0	[20, 25, 30, ..., 25]	[1.5, 1.4, 1.8, ..., 1.2]	**5.0**
2	12.0	23.0	[18, 22, 23, ..., 28]	[1.6, 1.3, 1.7, ..., 1.2]	**5.2**
3	11.5	22.4	[20, 21, 28, ..., 30]	[1.6, 1.6, 1.7, ..., 1.3]	**5.3**
n	...	...	...	...	...

The input data, as presented in Table 2.1, cannot be used directly for training a symbolic model, because traditional training methods can only handle scalar input variables and do not support vector data directly. Typically, features are extracted from those vectors, by using simple statistical measures such as mean and standard variation, or more sophisticated features, such as the linear trend of a vector or some form of similarity between two vectors. However, calculating all potentially useful features for all vectors can be unfeasible because it increases the number of features drastically, especially if considering features that require more than one vector, thus requiring all combinations of vectors.

Instead of performing feature engineering manually, the learning algorithm itself should find the proper extraction steps. For example, using the data from the artificial production plant, if the mean temperature is relevant for predicting the quality, the model itself should represent that fact. In some cases, it might be necessary to combine and modify vectors first before obtaining useful information, for example, while the spread of the temperature or the pressure is not important when considered separately, the spread of the variables combined might be. Those examples could be represented in a prediction model such as

$$Quality = \frac{1.5 \cdot MatA \cdot \text{mean}(Temp)}{MatB^2 + \text{stdev}(Press \cdot Temp)}. \tag{2.1}$$

While the presented model is simple—and most certainly wrong—it shows that the actual extraction step can also be performed by the symbolic model itself. The model might even come up with some extraction steps, that weren't obvious beforehand, and therefore, would have not been included via feature engineering. In essence, this eliminates manual feature engineering of vectors, and allows the model itself to describe how vector variables should be manipulated and combined with regular, scalar variables or other vector variables to describe the target variable.

As a general goal, we want to make symbolic regression more flexible towards different types of input data. We aim to reduce the need of manual feature engineering by allowing the models themselves to represent the necessary steps of extracting useful information. In this case we used time series measurements from a production plant as an example, however, the methodology also applies for any data of higher dimensions.

The remainder of the paper is structured as follows. Sect. 2.2 covers the state of the art regarding vectorial GP for symbolic regression and also grammar-based GP. Sect. 2.3 describes our method and discuss benefits and potential drawbacks compared to the state of the art. Sect. 2.4 explains our experimental setup and Sect. 2.5 contains the results and a detailed analysis for each presented benchmark instance. Finally, in Sect. 2.6, we summarize our findings and discuss further ideas and next steps.

2.2 State of the Art

In this section, we discuss the relevant state of the art for using vector data in symbolic regression. We also discuss grammar-based genetic programming and its benefits. Additionally, we also briefly mention other related topics such as feature engineering and deep learning.

2.2.1 Vectorial Genetic Programming

One way of using vector data in GP is using the data directly within the symbolic model via variables that represents vectors.

One of the first usages of vector data directly in GP was done by Holladay et al. using a stack-based GP language that is capable processing elements of different dimensions [11, 12]. Control structures, such as loops and branches, are used to operate on the vector data, to gradually convert values into lower dimensions. Additionally, more complex processing steps are also included to operate on signal data, such as a fast Fourier transform. Due to the nature of stack-based GP, the resulting models can be very complex and are also difficult to interpret.

Azzali et al. proposed a tree-based approach for vectorial GP, where both scalars and vectors can be used within the model, and operations on vectors are simply executed component wise [3, 4]. For instance, while the log of a scalar returns a scalar, the log of a vector returns a vector. Along common arithmetic operations, various aggregation functions are included to convert vectors into scalars, whereby scalars are simply treated as vectors of length one. This special case is necessary, because scalar and vector results can be mixed arbitrary. Lastly, many operators include additional parameters, controlling the range of an aggregation function, for instance. On the presented benchmarks, the presented tree-based vectorial approach shows very good results in comparison to traditional GP.

One example of a tree that also includes vector is shown in Fig. 2.2. Here, the variables $\boldsymbol{p}$ and $\boldsymbol{t}$ represent vectors, that are multiplied component-wise, before the maximum is obtained. Then, the maximum is added to the scalar variable w to produce the final model output.

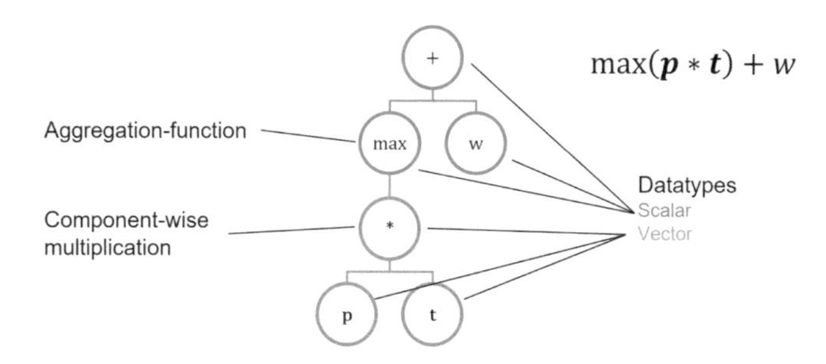

Fig. 2.2 During the evaluation of a symbolic model, each subtree can return a different type, depending on the type(s) of the input(s). In this case, the nodes for vector variables p and t yield vectors, therefore, their component-wise product yields a vector. The max symbol takes a vector as input but yields a scalar

The tree-based approach by Azzali et al. is very flexible, since all operations can be applied to both scalar and vector data.[1] However, this flexibility can also be a downside, for instance, when the final type of the prediction needs to be of a certain type. If the final prediction does not match the required type, a penalty is applied to guide GP towards the required output type. Another potential issue could also arise if an operation requires a specific type for one of its arguments, which would require some form of error handling if an invalid argument type is supplied. Therefore, having constraints that prevent invalid models would be beneficial.

2.2.2 *Grammar-Based Genetic Programming*

In the previous Sect. 2.2.1 on vectorial GP, we identified, that being able to restrict the output type of the model or the input types of certain operations would avoid certain problems that otherwise need to be handled via penalties or other error mechanisms. One way of restricting the model space for GP is grammar-based GP, which is briefly summarized in this section.

Early ideas of grammar-based GP were introduced by Montana with strongly-typed GP, where each tree-node is assigned a type and a dimension, e.g. a tree-node with INTEGER-3 always returns an integer vector of length 3 [17]. This way, strongly-typed GP can already handle data of different dimensionality, however, each function takes only arguments of a type with fixed dimensions, for instance, the operation "vector-add-3" can only add two vectors of length 3. This limitation was also addressed by Montana, by having generic operations that supported arbitrary dimensions, allowing strongly-typed GP to flexibly define the required input type of each argument of a function, and thus avoiding invalid models where functions cannot be called due to mismatched argument types. During the GP process, genetic

[1] Similar behavior could also be achieved by only using vectors and treating scalars as vectors of length one.

operators such as mutation and crossover must adhere to the specified constraints to only produce valid children. This is usually by maintaining internal data structures to query potentially valid inputs for a given function, so that crossover- and mutation points can be selected that will result in valid a offspring.

While strongly-typed GP handles constraints on input types of function arguments to avoid invalid models, sometimes there is a need to specify further constraints to avoid models that are valid in a technical sense, but still unwanted from a user's perspective. For instance, consider a function to calculate the power, where both base and exponent can be specified and are required to be real-valued—which can already be done via strongly-typed GP. To avoid overly complex models using this power function where the exponent could by any value calculated by a sub-tree, we want to limit the exponent argument to be a fixed, constant-value. However, strongly-typed GP could only enforce that the exponent is real-valued but not that the value must come from a constant. Therefore, an additional way of specifying and enforcing structural constraint on the symbolic model is necessary.

In grammar-based GP, the search space of allowed models is specified declaratively by a formal grammar [15], much like grammars for spoken languages defines how a correct sentence must be structured. Formal grammars are specified by terminal and non-terminal symbols (similar to tree-based GP), where the terminals are allowed symbols for the final sentence, which is a symbolic model in our case. The non-terminal symbols in the grammar define derivation rules, describing how the terminal and non-terminal symbols are allowed to nest, therefore describing the allowed structure of a model. This limits the search space for GP drastically, since many models are excluded beforehand. Like in strongly-typed GP, crossover, mutator and random creation of models must always adhere to the grammar.

In formal languages, grammars are often specified in textual form, such as the extended Backus-Naur form (EBFN).[2] Although the EBNF supports many powerful mechanisms for defining a grammar, for the examples in this paper we only need the definition of non-terminals (in pascal casing), terminals (in camel casing or enclosed in quotation marks) and alternatives (by separating the alternatives by a |). A simple grammar for symbolic regression could look like the following:

```
Start       = RealVal ;
RealVal     = Arithmetic | Power | Terminal ;
Arithmetic  = RealVal "+" RealVal | RealVal "-" RealVal
            | RealVal "*" RealVal | RealVal "/" RealVal ;
Power       = RealVal "^" constant ;
Terminal    = constant | variable ;
```

This grammar allows arbitrary arguments for arithmetic operators and for the base of the power (`"^"`) but limits the exponent to a constant. A model `3*a + b^2` would be valid according to this grammar, `a^b` would not. Note that for this simple grammar, there are no parentheses, and the order of operations is defined as semantic behavior by the derivation tree when creating the model.

[2] See ISO/IEC 14977

Symbols in the `Terminal` group will be terminals for the symbolic model and require some special handling. The `constant` terminal symbol represents any numeric constant, such as `"3.41"`. The `variable` terminal symbol can be any name of an input variable from the training data, for example `"x1"`.

2.2.3 *Feature Engineering and Feature Extraction*

Instead of using vector data directly as input features, one common practice is to extract scalar input features from the vectors beforehand, which can be simple statistics, or also complex relations identified by experts.

Interestingly, GP was used as feature engineering mechanism by various authors already. Harvey and Todd, for instance, used GP to create simple pipelines for feature extraction of sequential data [10]. Eads et al. used GP to extract features from time series data, that were then used in a support vector machine for prediction [7]. Likewise, other authors used GP for feature engineering from higher dimensional data, but did not use GP as a prediction model itself [2, 6, 19].

In this paper, we want to use GP as the prediction model where feature engineering is done as an included step. And since many authors already used GP successfully as feature extractor, their ideas serves as valuable inspiration on which aggregation functions were already successfully used in the past. Also, general feature extraction frameworks, such as *tsfresh*, serves as reference for potential aggregation functions [5].

2.2.4 *Deep Learning*

Neural networks are one of the few machine learning algorithms that directly support higher dimensional data, such as vector data or even higher-dimensional data representing images and videos [8]. Additionally, deep neural networks are often advertised as machine learning technique that eliminates any manual feature engineering, because feature engineering is already performed as part of the network and it's many layers. As a result, deep neural networks offers some powerful features that are not established well in GP yet. However, neural network lack the simplicity and interpretability of symbolic models, therefore, the knowledge on how features were extracted from the higher dimensional data, and how those effect a model's output is very difficult to assess.

There is an also interesting technical intersection between GP and neural networks regarding the representation of models. Most neural networks libraries, such as TensorFlow[3] or PyTorch,[4] use directed acyclic graphs of tensor operations for

[3] https://www.tensorflow.org

[4] https://pytorch.org

representing a neural network, similar to the graph representation of Cartesian GP [16]. Such tensor-graphs could also be used to represent symbolic models that contain scalar and vector data. The additional benefit of representing models as directed graphs is it's ability to automatically calculate the gradient of such a graph via automatic differentiation [9], which is essential for training neural networks. Being able to calculate the gradient of a symbolic model also allows optimizing the numerical constants of a given symbolic model via least squares, which is more efficient than tuning the coefficients via GP [13].

2.3 Grammar-Based Vectorial Genetic Programming

In this section, we present our method that extends classical tree-based genetic programming for symbolic regression, to also support vector variables along scalar variables. Our method combines the vectorial GP approach described in Sect. 2.2.1 and a grammar-based approach for symbolic regression described in Sect. 2.2.2. Fortunately, both vectorial GP and grammar-based GP can be implemented independently since they affect independent aspects of GP.

On the one hand, extending GP to be able to handle vector variables only affects the interpretation module of GP, which is responsible for evaluating a tree-model, by applying operations defined in the sub-trees onto its argument and propagating the results bottom-up until it reaches the root of the tree. On the other hand, the grammar limits the search space by restricting random creation, crossover and mutator to only create models that adhere to the specified grammar. Therefore, while the interpretation module should be as powerful as possible and allow evaluation of all valid models, the grammar is responsible of limiting the search space to models that are sensible and the users are interested in.

2.3.1 Vectorial Tree Interpretation

Compared to traditional GP, in vectorial GP the interpretation module has to handle a larger number of potential combinations of arguments and operations since there are different argument types. For instance, the interpretation module must be able to handle the log for both scalar values and also for vector values. In most cases, this can be done straightforward, by simply applying functions component-wise. In cases with multiple arguments, the applied operation depends on the types of the arguments, for instance, when calculating the sum of two arguments, there are four potential combinations of types for the two arguments: $s + s$, $s + \boldsymbol{v}$, $\boldsymbol{v} + s$ or $\boldsymbol{v} + \boldsymbol{v}$, where s represent a scalar argument and $\boldsymbol{v}$ a vector argument.

$s + s$ can be easily handled by simple adding the arguments. $\boldsymbol{v} + \boldsymbol{v}$ represents a component-wise addition of the two arguments, assuming the vectors have the same length. For the cases $s + \boldsymbol{v}$ and $\boldsymbol{v} + s$, where the types differ, we interpret the scalar

Table 2.2 Input and output types of symbols for most (non-aggregating) functions and operations

Function with a single input	
Input type	Output type
Scalar	Scalar
Vector	Vector

Function or operation with two inputs		
First input type	Second input type	Output type
Scalar	Scalar	Scalar
Scalar	Vector	Vector
Vector	Scalar	Vector
Vector	Vector	Vector

as vector, repeating its value to match the vector argument's length and then apply the operation component-wise. This behavior mimics the broadcasting rules of the popular Python library NumPy.[5] The presented examples can also be generalized to more then two arguments, meaning that if any argument is a vector, the result will be a vector, otherwise the result will be a scalar. Table 2.2 summarizes the type tables for functions with a single or two arguments.

In addition to the operations that are commonly available for traditional symbolic regression, vectorial GP also requires operations that are able to reduce a vector to a scalar value. In this paper, we refer to this operations as *aggregation functions*. Aggregation functions can be simple statistical measures, such as the mean or standard deviation, or more complex functions, such as the linear trend of a vector. Contrary to non-aggregating functions, aggregation functions always return a scalar.

Because the interpretation module for evaluating GP models should be as flexible as possible, aggregation functions must also handle the special case of aggregating a scalar. In such a case, each aggregation function will either return the scalar input itself or return a specific default value. For example, while the mean of a scalar will simply return the scalar, the standard deviation of a scalar will always return zero.

Similar to regular operations, aggregation functions can also have multiple arguments, for instance, when calculating the covariance of two vectors. Again, if an aggregation function with two parameter is called with a vector and a scalar argument, the scalar is expanded to match the vector's length. However, the output will always be a scalar. Table 2.3 summarizes the type tables for aggregation functions with a single or two arguments.

2.3.2 Vectorial Symbolic Regression Grammar

Although the interpretation module can handle arbitrary complex models, having GP search through this complex and large search space can be infeasible and result in models that are unwanted by the user. Therefore we want to enforce some rules on how vectors and scalars can interact via a grammar, as discussed in Sect. 2.2.2.

[5] https://numpy.org/doc/stable/user/theory.broadcasting.html

Table 2.3 Input and output types of symbols for aggregating functions

Aggregation function with a single input	
Input type	Output type
Scalar	Scalar
Vector	Scalar

Aggregation function with two inputs		
First input type	Second input type	Output type
Scalar	Scalar	Scalar
Scalar	Vector	Scalar
Vector	Scalar	Scalar
Vector	Vector	Scalar

One key aspect that we want to limit with a grammar is the output type of the model. While the interpretation module supports both scalars and vectors, the user usually has a specific output type in mind. For regression, a real-valued scalar is typically required, although a real-valued vector would be possible. However, for this paper, we assume that we want to obtain a symbolic model with a scalar output.

To avoid unnecessary aggregations of scalars, we also want to restrict aggregation functions to only allow vectors as arguments. Of course, there are many other application-specific restrictions possible, however, in this section we want to present a general purpose grammar for vectorial GP.

For defining a grammar for vectorial Symbolic Regression (SR), we start with a grammar for regular SR. As a first step, we group operations and functions to keep the grammar simple and clean, for instance, sin, cos and tan are put into a separate group that represents trigonometric functions. Each group is then represented by a non-terminal symbol in the grammar, where the alternatives are the different functions of this group.

All groups are then grouped again into an overall scalar group. This group, represented as the non-terminal symbol `Scalar`, represents any scalar produced via any operations that yields a scalar, and therefore can be used in the grammar whenever there are no restrictions for an input argument.

A typical grammar for SR can look like the following:

```
Start          = Scalar ;
Scalar         = Arithmetic | Trigonometric | Exponentiation
               | ... | Terminal ;
Arithmetic     = Scalar "+" Scalar | Scalar "-" Scalar
               | Scalar "*" Scalar | Scalar "/" Scalar ;
Trigonometric  = "sin(" Scalar ")" | "cos(" Scalar ")"
               | "tan(" Scalar ")" ;
Exponentiation = "power(" Scalar "," constant ")"
               | "root(" Scalar "," constant ")"
               | "log(" Scalar "," constant ")" ;
...
Terminal       = constant | variable ;
```

In the example grammar, most arguments can by any `Scalar` to allow unrestricted arguments. However, the second arguments of the functions within the `Exponentiation` group are limited to constants, to a void overly complex models, also also mentioned in Sect. 2.2.2.

For a vectorial grammar that handles both scalars and vectors, we use the existing grammar and extend the grammar by a `Vector` group that handles all operations that will return vectors. Both group look quite similar, since both can perform the same functions, however, all operations in the `Vector` group will be applied component wise over the vector arguments, as described earlier in Sect. 2.3.1. To distinguish the symbols within the `Vector` group from the symbols from the `Scalar` group, the vector groups are prefixed with `Vec`. For example, while the group `Trigonometric` operators on scalars, the group `VecTrigonometric` operates on vectors. The resulting grammar for vectorial SR can look like the following:

```
Start = Scalar ;

Scalar         = Arithmetic | Trigonometric | Exponentiation
               | ... | Aggregation |  Terminal ;
Arithmetic     = Scalar "+" Scalar | Scalar "-" Scalar
               | Scalar "*" Scalar | Scalar "/" Scalar ;
Trigonometric  = "sin(" Scalar ")" | "cos(" Scalar ")"
               | "tan(" Scalar ")" ;
Exponentiation = "power(" Scalar "," constant ")"
               | "root(" Scalar "," constant ")"
               | "log(" Scalar "," constant ")" ;
...
Aggregation    = Statistic | Distance ;
Statistic      = "mean(" Vector ")" | "std(" Vector ")" | ... ;
Distance       = "cov(" Vector "," Vector ")"
               | "euclidean(" Vector "," Vector ")" | ... ;
Terminal       = constant | variable ;

Vector            = VecArithmetic | VecTrigonometric
                  | VecExponentiation | ... | VecTerminal ;
VecArithmetic     = (Vector | Scalar) "+" (Vector | Scalar)
                  | (Vector | Scalar) "-" (Vector | Scalar)
                  | (Vector | Scalar) "*" (Vector | Scalar)
                  | (Vector | Scalar) "/" (Vector | Scalar) ;
VecTrigonometric  = "sin(" Vector ")" | "cos(" Vector ")"
                  | "tan(" Vector ")" ;


VecExponentiation = "power(" Vector "," constant ")"
                  | "root(" Vector "," constant ")"
                  | "log(" Vector "," constant ")" ;
...
VecTerminal       = vec-variable;
```

Within the new `Vector` group there are two key differences compared to the `Scalar` group. First, in the `VecArithmetic` group both `Vector` and `Scalar` are allowed as input, allowing, for instance, multiplication of a vector with a scalar.

Second, the `VecTerminal` only contains `vec-variable`, meaning that we do not allow vector-constants with this grammar.[6] The `vec-variable` also can only represent variables that are vectors.

Also the `Scalar` group differs and now contains a new group `Aggregation` that contains aggregation functions. The `Aggregation` group is a sub-group of the `Scalar` group because symbols from this group will return a scalar, although the arguments of the aggregation functions are `Vectors`. This ensures, that aggregation functions are only applied to vectors, and not to scalars.

Since the grammar still defines that the `Start` must be a `Scalar`, the final model of the output will always be a scalar. In case the model should predict a whole vector, this the start symbol can easily be changed to `Vector`.

We also want to note a minor design flaw in the presented grammar. Since the operations in `VecArithmetic` allow both vectors and scalars as argument types, it is valid that both arguments are scalars, resulting in the `VecArithmetic` group to return a scalar. Although the interpretation module can handle such a case, it should be avoided if possible.

One option would be to only allow vectors as arguments, however, that would disallow common operations such as multiplying a vector by a scalar. Another option would be to enforce that the first argument has to be a vector. This option would not limit the search space since the arguments could simply be switched for commutative operations, e.g. $s + \boldsymbol{v}$ could simply be changed to $\boldsymbol{v} + s$. Non-commutative operations, on the other hand, can first be converted to a commutative one by negating or inverting the second argument, i.e. $s - \boldsymbol{v}$ could be changed to $(-\boldsymbol{v}) + s$. A third option would also be to list all allowed combinations exhaustively, however this will result in a large number of alternatives, since this has to be done per operation.

2.4 Experiment Setup

In this section, we present the experimental setup for comparing vectorial GP to classical GP without vectors, where both use a grammar for limiting the search space. All experiments are conducted in HeuristicLab,[7] an open-source framework for heuristic optimization, that also includes grammar-based GP [14, 18].

To the authors best knowledge, there are no comprehensive benchmark datasets for regression that includes vector variables but requires a scalar target variable. Therefore, we created a new benchmark suite, where we included both scalar and vector variables. The benchmark functions are kept very simple, to allow detailed analysis off the core aspects of vectorial GP, without having to deal with issues when

[6] We avoided vector constants on purpose to avoid the problem of figuring out the correct vector length for a vector constant.

[7] https://dev.heuristiclab.com/

the training is very difficult. All benchmarks are listed in the Appendix, and the generated data for the experiment can be found online.[8]

Because traditional GP cannot use vectors as input variables directly, we cannot use the exact same benchmark instances. Instead, we transform the benchmark instances that contain vector variables into benchmark instances that only contain scalar variables, without changing the underlying data or equation for generating the target variable. As a result, we obtain three *variants* of a benchmark instance, one for vectorial GP and two for traditional GP:

Vector Variant: This variant includes both scalar and vector input features, and represent the "raw" version of the benchmark instance, as defined in the Appendix. This variant will be solved using grammar-based vectorial GP.

Pre-Aggregated Variant: This variant represents a traditional GP setup, where feature engineering was performed beforehand to convert vectors to scalars. For this experiment we used the following statistics: mean, median, standard deviation, variance, min and max. Due to the aggregation step, this variant only contains scalar input features. Also, the number of features is a higher than the vector variant, because each vector is aggregated into multiple features. Pre-aggregating the vectors into scalars will cause some degree of information loss, therefore, we assume that for some benchmarks, this variant will fall behind the vector variant where all information is still available.

Unrolled Variant: This variant of the benchmark represents a use case, where all information from the vector variables is preserved, but as individual scalar variables. For this, each vector variable is unrolled into multiple scalar variables, e.g. a 20-dimensional vector is unrolled into 20 separate scalar variables. Compared to the pre-aggregated variant, this variant does not loose any information to feature engineering, offering the same raw information as the vector variant. However, the number of variables can be substantially larger because each vector is split into many scalar variables. Also, to make use of all the variables, the resulting symbolic model has to become much larger and therefore, is computationally more expensive to train and also more difficult to interpret due to its size. This variant is mainly included for reference and does not represents a practicable solution when dealing with vector data.

Note, that all three variants are based on the same raw data. This means that the vector variant and the unrolled variant contain exactly the same data, but arranged into different variables. The pre-aggregated variant contains different data due to the aggregation step, but the raw data is exactly the same as the other variants.

In addition to the three variants of the benchmarks, the benchmarks are also split into two groups, according to their difficulty.

- Group A contains simple benchmarks where only direct aggregations of vector-variables are necessary. Those simple benchmarks should pose no difficulty for the vector variant and the pre-aggregated variant. For the unrolled variant, however,

[8] https://dev.heuristiclab.com/wiki/AdditionalMaterial#GPTP2021

Table 2.4 Algorithm configuration

Population Size	1000
Elites	1
Selection	Gender-specific [a]
Crossover	Subtree swapping
Crossover Probability	100 %
Mutation	Multiple [b]
Mutation Probability	20 %
Comparison Factor	1.0
Max Generations	200
Max Selection Pressure	200
Max Tree Length	30 / 60 [c]
Training-Test Split	75 % / 25 %
Optimization Criterion	Coefficient of determination ($R^2 = \frac{\text{cov}(y,\hat{y})^2}{\text{var}(y)\text{var}(\hat{y})}$)

[a] Proportional & random
[b] Change node type, Full tree shaker, One point shaker, Remove branch, Replace branch
[c] Larger trees are allowed for the unrolled variant

aggregating a vector manually is considerable more difficult, especially for non-trivial aggregations such as the variance of a vector. Thus, we expect that the unrolled variant will already fail on benchmarks in this group.

- Group B contains benchmarks, where vector variables interact with other vector variables before aggregation. While the vector variant should be able to handle those cases, we expect that the pre-aggregated variant won't always be able to obtain the same information because some information is already lost while feature engineering.

For training we use Offspring Selection GP (OS-GP), since the algorithm is more robust than standard GP [1]. The configuration of OS-GP is listed in Table 2.4. Table 2.5 lists the allowed terminal and non-terminal set, which are defined via a grammar, as described in Sect. 2.3.2.

Since the benchmarks are very simple, GP should be able to solve them robustly if the configuration is appropriate. Therefore, we have deliberately chosen a relatively large population size to avoid potential unlucky runs with a suboptimal initial population, that then struggles to reintroduce necessary building blocks via mutation. Likewise, we narrowed the search space down by only include relevant functions in the grammar that are necessary to solve the benchmarks. Or course, GP should be able to solve the instance with smaller population sizes and larger search spaces, but we want each GP run to reliably solve the instance, if it is able to.

Table 2.5 Allowed terminal and non-terminal symbols for the experiment. Symbols dealing with vectors are only present for the vector variant

Symbol-Group	Symbols	Types
Terminals	scalar constant, scalar variable, vector variable*	
Arithmetics	$+, -, \cdot, /$	Scalar and Vector*
Exponentiation	x^2, $\sqrt{x}$	Scalar and Vector*
Aggregations*	mean, median, std, var, min, max	Vector only

[a] Only for the vector variant

2.5 Results

For the results, each dataset contains 1000 rows and each vector contains 20 elements. Each GP run was repeated 50 times for each variant and instance, with the aggregated results listed in Table 2.6.

Because the benchmark instances are defined without noise, they can be solved exactly, i.e. with $R^2 = 1$ or $NMSE = 0$. Therefore, we also analyzed the success rates of each of each benchmark instance by variant. To this end, we define a success threshold of a maximum $NMSE$ of 10^{-4}, meaning that we consider a benchmark solved, if the $NMSE$ on the test partition is less than 10^{-4}. On the relation that $NMSE \approx 1 - R^2$, this means we consider a benchmark as solved, if the unexplained variance of the solution is smaller than 0.01%. Table 2.7 shows the number of successful runs and the success rate for all instances per variant.

The results show that the vector variant was capable of solving all benchmark instances from benchmark group A and B. The pre-aggregated variant was also able to reliably solve all benchmarks of group A, however, was only able to solve benchmark 01 in benchmark group B. The unrolled variant was only sometimes able to solve some simple benchmarks of group A.

In the following, we analyze each benchmark instance in more detail, explaining why each instance could or could not be solved the different variants. First, we discusses the simpler benchmark group A where the vector and the pre-aggregated variant was able to solve all instances, while the unrolled variant struggled already. Second, we discusses benchmark group B where interactions between vector variables are required before aggregation, demonstrating the limits of pre-aggregating vector data into scalar variables. For each instance the function is listed along the success-rates on the right (V=vector, A=pre-aggregated, U=unrolled).

Table 2.6 Mean ± standard deviation of the coefficient of determination R^2 and normalized mean squared error ($NMSE(y, \hat{y}) = \frac{MSE(y,\hat{y})}{\mathrm{var}(y)}$) on the test-partition for each instance, grouped by variant

Instance	Vector	Pre-Aggregated	Unrolled	Vector	Pre-Aggregated	Unrolled
		R^2			$NMSE$	
test_A_01	1.0000 ± 0.00	1.0000 ± 0.00	0.9954 ± 0.00	0.0000 ± 0.00	0.0000 ± 0.00	0.0046 ± 0.00
test_A_02	1.0000 ± 0.00	1.0000 ± 0.00	0.3360 ± 0.12	0.0000 ± 0.00	0.0000 ± 0.00	0.6319 ± 0.07
test_A_03	1.0000 ± 0.00	1.0000 ± 0.00	0.9971 ± 0.00	0.0000 ± 0.00	0.0000 ± 0.00	0.0029 ± 0.00
test_A_04	0.9998 ± 0.00	0.9999 ± 0.00	0.4747 ± 0.10	0.0002 ± 0.00	0.0001 ± 0.00	0.5208 ± 0.09
test_B_01	1.0000 ± 0.00	1.0000 ± 0.00	0.9876 ± 0.00	0.0000 ± 0.00	0.0000 ± 0.00	0.0126 ± 0.00
test_B_02	1.0000 ± 0.00	0.7557 ± 0.27	0.9401 ± 0.01	0.0000 ± 0.00	0.1921 ± 0.18	0.0614 ± 0.01
test_B_03	1.0000 ± 0.00	0.9190 ± 0.04	0.6830 ± 0.15	0.0000 ± 0.00	0.0833 ± 0.05	0.3055 ± 0.14
test_B_04	1.0000 ± 0.00	0.9176 ± 0.01	0.7266 ± 0.06	0.0000 ± 0.00	0.0810 ± 0.01	0.2838 ± 0.08
test_B_05	1.0000 ± 0.00	0.9773 ± 0.02	0.7800 ± 0.12	0.0000 ± 0.00	0.0257 ± 0.03	0.2168 ± 0.12
test_B_06	0.9998 ± 0.00	0.9596 ± 0.06	0.4003 ± 0.27	0.0004 ± 0.00	0.0373 ± 0.01	0.6359 ± 0.23

Table 2.7 Number of successful runs, grouped by variant. A benchmark is solved successfully, if the *NMSE* on the test-partition is less than 10^{-4}

Instance	Vector	Pre-Aggregated	Unrolled
test_A_01	50/50 (100%)	50/50 (100%)	6/50 (12%)
test_A_02	50/50 (100%)	50/50 (100%)	0/50 (0%)
test_A_03	50/50 (100%)	50/50 (100%)	6/50 (12%)
test_A_04	49/50 (98%)	49/50 (98%)	0/50 (0%)
test_B_01	50/50 (100%)	50/50 (100%)	0/50 (0%)
test_B_02	50/50 (100%)	0/50 (0%)	0/50 (0%)
test_B_03	50/50 (100%)	0/50 (0%)	0/50 (0%)
test_B_04	50/50 (100%)	0/50 (0%)	0/50 (0%)
test_B_05	49/50 (98%)	0/50 (0%)	0/50 (0%)
test_B_06	47/50 (94%)	0/50 (0%)	0/50 (0%)

2.5.1 Analysis Benchmarks Group A

test_A_01 $y = 2.5 \cdot \text{mean}(\boldsymbol{v}_1)$
V: 100% A: 100% U: 12%

This very simple benchmark is designed as a proof of concept for vectorial GP, which could be solved by the vector variant and the pre-aggregated variant easily. However, the unrolled variant has to perform the mean aggregation manually by summing all individual variables and dividing by the number of variables. Since for the benchmarks the vectors contain 20 values, 20 individual variables have to be summed, which is already difficult for standard GP, hence the low success rate of 12%.

test_A_02 $y = \text{std}(\boldsymbol{v}_1) - 2$
V: 100% A: 100% U: 0%
Complexity-wise, this instance is similar to the previous one, but using the standard deviation for aggregation. This is still easily solvable with the vector variant by using the standard deviation aggregation function or with the pre-aggregated variant by using the already pre-calculated standard deviation. However, the equation for calculating the standard deviation of 20 values manually is quite challenging for GP. For this calculation, the deviation of each value to the mean is required first, meaning that the mean of the 20 values has to be calculated beforehand, and since the tree-based models cannot store the mean for multiple use, it has to be recalculated 20 times. Therefore, the unrolled variant was not able to solve this benchmark.

test_A_03 $y = 2.5 \cdot x_1 + \text{mean}(\boldsymbol{v}_1) + 2.0$
V: 100% A: 100% U: 12%

This instance differs to test_A_01 only in using an additional scalar variable x_1. As a result, both vector and pre-aggregated variable can easily solve the instance, while the unrolled variant only managed to solve it in 12% of the runs.

test_A_04 $y = x_1 \cdot \mathrm{var}(\boldsymbol{v}_1)/3.0 - 3.0 \cdot \mathrm{mean}(\boldsymbol{v}_2)/x_2$
V: 98% A: 98% U: 0%
This instance, uses two vector variables and requires two different aggregations, meaning that this instance is a bit harder to solve due to additional variables and operations. However, both vector and pre-aggregated variant were both able to solve the instance almost every time. Similar to test_A_02, the unrolled variant was not able to calculate the variance manually, thus, was unable to solve this instance.

2.5.2 *Analysis Benchmarks Group B*

test_B_01 $y = x_1 \cdot \mathrm{mean}(\boldsymbol{v_1} + \boldsymbol{v_2})$
V: 100% A: 100% U: 0%
This instance requires a mean aggregation of the sum of two vectors. Because, the expected value of the sum of two random variables is the sum of the expected values, i.e. $E[X + Y] = E[X] + E[Y]$, this instance can both be solved by the vector variant and the aggregated variant. The vector variant can either calculate the sum of the vectors first, and then calculate the mean, or calculate the mean of the vector separately and them sum the means. The pre-aggregated variant always can simply add the pre-calculated means of the vectors. With sufficiently sized trees, the unrolled variant should also be able to solve this, however, would require considerable more resources and would result in a very large model.

test_B_02 $y = x_1 \cdot \mathrm{mean}(\boldsymbol{v}_1 \cdot \boldsymbol{v}_2)$
V: 100% A: 0% U: 0%
This instance is similar to the previous test_B_01, however the mean of the product of two vectors is used. In general, the expected value of the product of two random variables is not equal the product of the expected value, i.e. $E[X \cdot Y] \neq E[X] \cdot E[Y]$, thus, the pre-aggregated variant is not able to solve this instance completely because multiplying the pre-calculated means does not yield the correct result. In cases where the two vectors are completely independent from each other, the expected value of the product would be equal to the product of the expected value. Although the data for the benchmarks is generated randomly, there is still some very small, random correlation, i.e. $|Cov[X, Y]| > 0$, therefore the instance could not be solved according to the defined success-threshold of $NMSE < 10^{-4}$. For real-world application, however, it is to be expected that measured signals of a system are connected and therefore will rarely be complete independent from each other, meaning that pre-aggregating vector data into scalars would not be able to maintain the relevant information to solve such a case.

test_B_03 $y = x_1 \cdot \mathrm{std}(\boldsymbol{v}_1 + \boldsymbol{v}_2)$
V: 100% A: 0% U: 0%

Since the standard deviation of the sum of two variables is not equal the sum of the standard deviations, i.e. $\mathrm{Var}[X+Y] = \mathrm{Var}[X] + 2\mathrm{Cov}[X, Y] + \mathrm{Var}[Y]$, the pre-aggregated variant could not solve this instance because the covariance was not one of the calculated features. Additional experiments have shown, that when the covariance of $\boldsymbol{v_1}$ and $\boldsymbol{v_2}$ is present, the aggregated variant was able to solve this instance robustly. However, when feature extraction is performed on vector data, covariances for all combinations of vector variables is rarely done, therefore we assume, that such a feature is typically not present in the training data.

test_B_04 $y = x_1 \cdot \mathrm{std}(\boldsymbol{v}_1 \cdot \boldsymbol{v}_2)$
V: 100% A: 0% U: 0%
Similar to test_B_02, the standard deviation of the product of two random variables can only be calculated if the random variables are independent. In cases where the vectors are completely independent, the aggregated variant could be able to discover that $\mathrm{Var}[X \cdot Y] = \mathrm{Var}[X] \cdot \mathrm{Var}[Y] + \mathrm{Var}[X] \cdot (E[Y])^2 + \mathrm{Var}[Y] \cdot (E[X])^2$. However, as discussed earlier, in real-world scenarios, it is to be expected that measured signals would typically be related to a certain extend, thus, the raw data is usually required to correctly calculate the standard deviation of the product.

test_B_05 $y = x_1 \cdot \mathrm{mean}((\boldsymbol{v}_1 + 2\boldsymbol{v}_3)/(0.5\boldsymbol{v}_2))$
V: 98% A: 0% U: 0%
This instance combines test_B_01 and test_B_02, showing that the vector variant can also solve instances where multiple combinations of different vectors are required before aggregation. Again, pre-aggregated and unrolled variant were not able to solve this instance, while the vector variant was able to solve it robustly.

test_B_06 $y = x_1 \cdot \mathrm{std}((\boldsymbol{v}_1 + 2\boldsymbol{v}_3)/(0.5\boldsymbol{v}_2))$
V: 94% A: 0% U: 0%
Similar to test_B_05, this instance is a combination of previous instance, but uses the standard deviation for aggregation. The vector variant was also able to solve this instance in almost all of the cases.

2.6 Discussion and Next Steps

The benchmarks and experimental setup presented in Sect. 2.4 were designed to show limitations of traditional GP when working with vector data. According to the results in Sect. 2.5, the first benchmark group A already showed that using vector data by simply unrolling the vector into multiple, scalar variables is not effective. First, it requires that the vectors have a fixed length, otherwise the number of variables would change. Second, even simple aggregations of medium size vector lengths would require symbolic models of considerable sizes to accommodate high number of variables. A common solution is to pre-aggregate vectorial data via a feature engineering step, and include promising key figures of the vectors in the training data. The results show, that pre-aggregating the data works well in cases where vectors are directly aggregated without any previous interaction with other vector variables.

The second benchmark group B contained instances where vector variables have to be combined with other vectors before aggregating them. Pre-aggregating vectorial data, however, is done on vectors directly, thus, some information loss is occurring. Therefore, when interaction of vector variables are required before aggregation, the required information is not present in the pre-aggregated variables, and the relevant information for performing the interaction is already lost. In such cases, only the vector variant was able to perform the necessary interactions prior to aggregation, since the data is still available in the form of vector variables.

The benchmarks in this paper were kept very simple on purpose and also have some restrictions concerning the vector lengths, which simplified the benchmarks. The first restrictions concerns the aggregation range of the vectors, meaning that aggregation was always applied on a whole vector. In practice, it could be very interesting to calculate the aggregation only on a subset of a vector. For instance, in real-world cases, the last minutes of a process could have a high impact, whereas measurements that where longer in the past have no influence. In this case, aggregation should only be calculated based on the parts of the vector that is most important. Of course, determining the optimal sub-section of the vector to aggregate is not known in advanced, and therefore, should be determined by GP itself. Aggregating only a subset of a vector is also a unique feature of vectorial GP that standard GP would have difficulty with, since aggregating on a subset would require that the original vector data is still present in the training data.

A second restrictions concerns the lengths of the vectors, which where constant for all presented benchmarks. However, in a typical production scenarios where vector data represent time series of sensor data, the lengths could differ per product, depending on the production time. In this case, all vectors of the sample have the same length and, therefore, would not cause an immediate issue for component-wise operations since they still operate on vectors of the same length. If the vector lengths vary within a single product, i.e. different vector variables have different lengths per observation, then additional mechanisms would be necessary to handle operations on vectors of different lengths. Such mechanisms could include resampling the vectors to compatible lengths, cutting the longer one, or filling up the shorter one. Regardless of the mechanism, benchmarks where the vector lengths differ would be an interesting addition to study the behavior and applicability of vectorial GP.

Another way forward could be a general extension for GP towards input types of higher dimensions. While we only covered vectors in this paper, the presented methods could also be extended towards matrices and input types of higher dimensions. Aggregation functions could gradually reduce dimensions, until the desired dimensionality is reached.

Acknowledgements This work was carried out within the Dissertationsprogramm der Fachhochschule OÖ #875441 *Vektor-basierte Genetische Programmierung für Symbolische Regression und Klassifikation mit Zeitreihen (SymRegZeit)*, funded by the Austrian Research Promotion Agency FFG. The authors also gratefully acknowledge support by the Christian Doppler Research Association and the Federal Ministry of Digital and Economic Affairs within the *Josef Ressel Centre for Symbolic Regression.*

Appendix

Below, the equations to generate the target variable for each benchmark are listed. Each scalar variable x_i is defined by a uniform distribution with lower and upper bound, denoted by $\mathcal{U}(\text{lower}, \text{upper})$. Each vector variable $\boldsymbol{v_i}$ is defined by two hidden, scalar variables for defining the mean and standard deviation of the vector. These hidden variables are also defined and obtained via uniform distributions, like a regular scalar variable. Then, based on the hidden mean and standard deviation for each vector variable, we define the final uniform distribution where the values for each individual vector is sampled from, where the upper and lower bound are $\mu \pm \sqrt{12}\sigma/2$. In this case, we denote the vector variable by $\mathcal{U}(\mu = [\text{lower mean, uper mean}], \sigma = [\text{lower std dev, upper std dev}]; \text{length})$. This step ensures that the mean and standard deviation of each vector is also randomized.

$$\begin{aligned}
\text{test_A_01} \quad & y = 2.5 \cdot \text{mean}(\boldsymbol{v_1}) \\
& \boldsymbol{v_1} \sim \mathcal{U}(\mu = [4, 8], \sigma = [2, 4]; 20) \\
\text{test_A_02} \quad & y = 2.5 \cdot \text{mean}(\boldsymbol{v_1}) \\
& \boldsymbol{v_1} \sim \mathcal{U}(\mu = [4, 8], \sigma = [2, 4]; 20) \\
\text{test_A_03} \quad & y = 2.5 \cdot x_1 + \text{mean}(\boldsymbol{v_1}) + 2.0 \\
& x_1 \sim \mathcal{U}(1, 4) \\
& \boldsymbol{v_1} \sim \mathcal{U}(\mu = [4, 8], \sigma = [2, 4]; 20) \\
\text{test_A_04} \quad & y = x_1 \cdot \text{var}(\boldsymbol{v_1})/3.0 - 3.0 \cdot \text{mean}(\boldsymbol{v_2})/x_2 \\
& x_1 \sim \mathcal{U}(1, 4) \\
& x_2 \sim \mathcal{U}(-8, -4) \\
& \boldsymbol{v_1} \sim \mathcal{U}(\mu = [4, 8], \sigma = [2, 4]; 20) \\
& \boldsymbol{v_2} \sim \mathcal{U}(\mu = [10, 20], \sigma = [4, 8]; 20) \\
\text{test_B_01} \quad & y = x_1 \cdot \text{mean}(\boldsymbol{v_1} + \boldsymbol{v_2}) \\
& x_1 \sim \mathcal{U}(1, 4) \\
& x_2 \sim \mathcal{U}(-8, -4) \\
& \boldsymbol{v_1} \sim \mathcal{U}(\mu = [4, 8], \sigma = [2, 4]; 20) \\
& \boldsymbol{v_2} \sim \mathcal{U}(\mu = [10, 20], \sigma = [4, 6]; 20) \\
\text{test_B_02} \quad & y = x_1 \cdot \text{mean}(\boldsymbol{v_1} \cdot \boldsymbol{v_2}) \\
& x_1 \sim \mathcal{U}(1, 4) \\
& x_2 \sim \mathcal{U}(-8, -4) \\
& \boldsymbol{v_1} \sim \mathcal{U}(\mu = [4, 8], \sigma = [2, 4]; 20) \\
& \boldsymbol{v_2} \sim \mathcal{U}(\mu = [10, 20], \sigma = [4, 6]; 20) \\
\text{test_B_03} \quad & y = x_1 \cdot \text{std}(\boldsymbol{v_1} + \boldsymbol{v_2})
\end{aligned}$$

$$\begin{aligned}
& x_1 \sim \mathcal{U}(1, 4) \\
& x_2 \sim \mathcal{U}(-8, -4) \\
& \boldsymbol{v_1} \sim \mathcal{U}(\mu = [4, 8], \sigma = [2, 4]; 20) \\
& \boldsymbol{v_2} \sim \mathcal{U}(\mu = [10, 20], \sigma = [4, 6]; 20) \\
\text{test_B_04} \quad & y = x_1 \cdot \text{std}(\boldsymbol{v_1} \cdot \boldsymbol{v_2}) \\
& x_1 \sim \mathcal{U}(1, 4) \\
& x_2 \sim \mathcal{U}(-8, -4) \\
& \boldsymbol{v_1} \sim \mathcal{U}(\mu = [4, 8], \sigma = [2, 4]; 20) \\
& \boldsymbol{v_2} \sim \mathcal{U}(\mu = [10, 20], \sigma = [4, 6]; 20) \\
\text{test_B_05} \quad & y = x_1 \cdot \text{mean}((\boldsymbol{v_1} + 2\boldsymbol{v_3})/(0.5\boldsymbol{v_2})) \\
& x_1 \sim \mathcal{U}(1, 4) \\
& x_2 \sim \mathcal{U}(-8, -4) \\
& \boldsymbol{v_1} \sim \mathcal{U}(\mu = [4, 8], \sigma = [2, 4]; 20) \\
& \boldsymbol{v_2} \sim \mathcal{U}(\mu = [10, 20], \sigma = [4, 6]; 20) \\
& \boldsymbol{v_3} \sim \mathcal{U}(\mu = [2, 4], \sigma = [0.05, 0.15]; 20) \\
\text{test_B_06} \quad & y = x_1 \cdot \text{std}((\boldsymbol{v_1} + 2\boldsymbol{v_3})/(0.5\boldsymbol{v_2})) \\
& x_1 \sim \mathcal{U}(1, 4) \\
& x_2 \sim \mathcal{U}(-8, -4) \\
& \boldsymbol{v_1} \sim \mathcal{U}(\mu = [4, 8], \sigma = [2, 4]; 20) \\
& \boldsymbol{v_2} \sim \mathcal{U}(\mu = [10, 20], \sigma = [4, 6]; 20) \\
& \boldsymbol{v_3} \sim \mathcal{U}(\mu = [2, 4], \sigma = [0.05, 0.15]; 20)
\end{aligned}$$

References

1. Affenzeller, M., Wagner, S.: Offspring selection: a new self-adaptive selection scheme for genetic algorithms. In: Adaptive and Natural Computing Algorithms, pp. 218–221 (2005)
2. Alfaro-Cid, E., Sharman, K., Esparica-Alcázar, A.L.: Genetic programming and serial processing for time series classification. Evol. Comput. (2013)
3. Azzali, I., Vanneschi, L., Bakurov, I., Silva, S., Ivaldi, M., Giacobini, M.: Towards the use of vector based GP to predict physiological time series. Appl. Soft Comput. J. **89**, 106097 (2020)
4. Azzali, I., Vanneschi, L., Silva, S., Bakurov, I., Giacobini, M.: A vectorial approach to genetic programming. EuroGP 213–227 (2019)
5. Christ, M., Braun, N., Neuffer, J., Kempa-Liehr, A.W..: Time series feature extraction on basis of scalable hypothesis tests (tsfresh A Python package). Neurocomputing **307**, 72–77 (2018)
6. De Falco, I., Della Cioppa, A., Tarantino, E.: A genetic programming system for time series prediction and its application to El Niño forecast. In: Soft Computing: Methodologies and Applications (2005)
7. Eads, D.R., Hill, D., Davis, S., Perkins, S.J., Ma, J., Porter, R.B., Theiler, J.P: Genetic algorithms and support vector machines for time series classification. In: Applications and Science of Neural Networks, Fuzzy Systems, and Evolutionary Computation V, vol. 4787, p. 74 (2002)
8. Goodfellow, I., Bengio, Y., Courville, A.: Deep Learning (2016)
9. Baydin, A.G., Pearlmutter, B.A., Radul, A.A., Siskind, J.M.: Automatic differentiation in machine learning: a survey. J. Mach. Learn. Res. **18**, 1–43 (2018)
10. Harvey, Dustin Y., Todd, Michael D.: Automated feature design for numeric sequence classification by genetic programming. IEEE Trans. Evol. Comput. **19**(4), 474–489 (2015)

11. Holladay, K.L., Robbins, K.A.: Evolution of signal processing algorithms using vector based genetic programming. In: 2007 15th International Conference on Digital Signal Processing, DSP 2007, pp. 503–506 (2007)
12. Holladay, K., Robbins, K., Von Ronne, J.: FIFTH: a stack based GP language for vector processing. In: Lecture Notes in Computer Science (including subseries Lecture Notes in Artificial Intelligence and Lecture Notes in Bioinformatics), 4445 LNCS(December), pp. 102–113 (2007)
13. Kommenda, M., Affenzeller, M., Kronberger, G., Winkler, S.M.: Nonlinear least squares optimization of constants in symbolic regression. In: Lecture Notes in Computer Science (including subseries Lecture Notes in Artificial Intelligence and Lecture Notes in Bioinformatics), 8111 LNCS(PART 1), pp. 420–427 (2013)
14. Kommenda, M., Kronberger, G., Wagner, S., Winkler, S., Affenzeller, M.: On the architecture and implementation of tree-based genetic programming in HeuristicLab. In: GECCO'12—Proceedings of the 14th International Conference on Genetic and Evolutionary Computation Companion, pp. 101–108 (2012)
15. McKay, R.I., Hoai, N.X., Whigham, P.A., Shan, Y., O'neill, M.: Grammar-based genetic programming: a survey. Genet. Program. Evolvable Mach. **11**(3–4), 365–396 (2010)
16. Miller, J.F., Harding, S.L.: Cartesian genetic programming. In: European Conference on Genetic Programming, pp. 121–132 (2000)
17. Montana DJ (1995) Strongly typed genetic programming. Evol. Comput. 1 **3**(2), 199–230 (1995)
18. Wagner, S., Kronberger, G., Beham, A., Kommenda, M., Scheibenpflug, A., Pitzer, E., Vonolfen, S., Kofler, M., Winkler, S., Dorfer, V., Affenzeller, M.: Architecture and design of the heuristiclab optimization environment 197–261 (2014)
19. Xie, F., Song, A., Ciesielski, V.: Event detection in time series by genetic programming. In: 2012 IEEE Congress on Evolutionary Computation, CEC 2012, pp. 1–8 (2012)

Chapter 3
Grammatical Evolution Mapping for Semantically-Constrained Genetic Programming

Alcides Fonseca, Paulo Santos, Guilherme Espada, and Sara Silva

Abstract Search-Based Software Engineering problems frequently have semantic constraints that can be used to deterministically restrict what type of programs can be generated, improving the performance of Genetic Programming. Strongly-Typed and Grammar-Guided Genetic Programming are two examples of using domain-knowledge to improve performance of Genetic Programming by preventing solutions that are known to be invalid from ever being added to the population. However, the restrictions in real world challenges like program synthesis, automated program repair or test generation are more complex than what context-free grammars or simple types can express. We address these limitations with examples, and discuss the process of efficiently generating individuals in the context of Christiansen Grammatical Evolution and Refined-Typed Genetic Programming. We present three new approaches for the population initialization procedure of semantically constrained GP that are more efficient and promote more diversity than traditional Grammatical Evolution.

3.1 Introduction

The Software Engineering practice includes solving high-complexity challenges that are traditionally done manually. However, researchers have tried to automate several phases of this process, resulting in the field of Search-Based Software Engineering

A. Fonseca (✉) · P. Santos · G. Espada · S. Silva
LASIGE, Departamento de Informática da Faculdade de Ciências da Universidade de Lisboa, Lisbon, Portugal
e-mail: alcides@ciencias.ulisboa.pt

P. Santos
e-mail: pacsanto@ciencias.ulisboa.pts

G. Espada
e-mail: gjespada@ciencias.ulisboa.pt

S. Silva
e-mail: sara@ciencias.ulisboa.pt

W. Banzhaf et al. (eds.), *Genetic Programming Theory and Practice XVIII*,
Genetic and Evolutionary Computation,
https://doi.org/10.1007/978-981-16-8113-4_3

(SBSE). Requirements engineering, specification mining, program synthesis, and test generation are examples of complex tasks that have been tackled using evolutionary algorithms.

SBSE is not the only approach to solve complex problems, though. Depending on the phase we are targeting, there are deterministic algorithms that are useful for solving the problem. However, deterministic algorithms are typically limited to a well-behaved restricted subset of programming languages, and often do not scale to more than two or three classes.

Taking Automated Program Repair as an example, there are approaches that perform a deterministic search following a set of rules (e.g., SemFix [25]), and there are approaches that rely on Genetic Programming [27] (GP) to find patches that pass all the tests (e.g., GenProg [11]). However, neither approach has been completely successful [6], and hybrid approaches like ARJA [39] are being proposed. ARJA uses GP, but encodes program information as rules in the genetic operators, avoiding parts of the search space that would not be viable.

This Chapter addresses the challenge of combining domain-knowledge constraints on the representation of the solutions with an efficient and generic evolutionary search procedure.

- Section 3.2 discusses several domains where restricting the tree construction with semantic information is useful.
- Section 3.3 presents existing approaches for encoding semantic constraints in GP.
- Section 3.4 presents our type-aware approach, compared to the existing generate-and-validate.
- Section 3.5 discusses possible individual representations for this problem.
- Section 3.6 presents the dynamic techniques for efficiently combining static and dynamic validation of trees.
- Section 3.7 reports on our empirical evaluation, applied to program synthesis.
- Section 3.8 draws the final conclusions.

3.2 Software Engineering Applications of Semantically–Constrained GP

SBSE covers a wide spectrum of the Software Development Life Cycle and related artifacts. While natural language can be converted to a more formal language (e.g., using the approach taken by Alhroob et al. [1]), this work focus on formal languages as they provide structure and information that can be used in reducing the search space. The idea of embedding constraints in the program representation is popular in the following application domains.

3.2.1 Automated Program Repair

Tools for automated program repair often fall in one these two categories: search-based and logic-based. Search-based approaches try to apply random template and ingredient-based mutations on the program components identified by Fault Localization tools. GenProg [11] was one of earliest approaches in this family.

On the other hand, SemFix [25] is an example of a logic-based tool that is deterministic in nature. Controlled symbolic test execution evidences the components of the program that fail to meet conditions required by the tests. Program synthesis is then used to synthesize a new, correct-by-construction, component to replace the buggy one. However, deterministic symbolic techniques are limited by the type of code supported by symbolic execution and are not better than search-based techniques [6]. The ideal solution would be to use search-based techniques constrained by the information recorded via semantic analysis, but there are challenges in efficiently including that information in search operators. For instance, ARJA [39] tries to address these, but the applied rules are only a small subset of the potential rules that can be applied, thus having a limited, albeit useful, benefit.

3.2.2 Automated Test Generation

Tools for automated test generation follow the same pattern: there are tools that rely mostly on evolutionary algorithms to find tests that maximize coverage (e.g., EvoSuite [9]) and tools that rely solely on symbolic execution to obtain deterministic tests that maximize branching conditions (e.g., ATGen [22]). Suggested hybrid approaches [10] rely on dynamic symbolic execution in fitness evaluation to guide search, with an evaluation overhead in execution time (which has led the authors to perform dynamic symbolic execution only once every 30 s). Recently, a similar approach with no overhead during population evaluation has been presented that also encodes semantic information into the program generation phase [33]. Another related approach has been to use the logical specification of a function to restrict the type of operators that can be applied to an S-expression [34].

3.2.3 Program Synthesis

Genetic Programming is, in itself, Program Synthesis. The goal of GP is to create a program that maximizes a given dynamic semantic specification. Alternative approaches rely on SMT/SAT solvers to more deterministically obtain the exact expression that conforms to a logical specification (popular in Program Sketching [35] and in Higher-Order functional programs [13]). While SMT-based approaches are orders of magnitude faster, they are limited to statically verifiable logical conditions, while GP can use as a fitness function any executable code. In fact, GP is also capable of solving the sketch-based problems [4]. Recently, a new approach has been proposed that integrates SMT solvers in the evolutionary search, in the context of Program Synthesis [7], which will be the focus of this Chapter. Another advantage of using GP with constraints is that it can be used to increase the readability of the generated programs [5].

3.3 Semantic Constraints in GP

While Standard GP imposes no restrictions on the representation of individuals, other than distinguishing terminals from non-terminals, more advanced alternatives have been proposed.

3.3.1 Strongly-Typed GP (STGP)

Montana has proposed the use of simple types to restrict the generation of program trees [24]. This is the first step in encoding program properties in the tree representation, which excludes most of the (irrelevant) search space. In fact, the more restrictions there are on the program tree, the more efficient search will be. As an example, STGP has been used to generate unit tests for object-oriented software [38]. STGP has been extended with polymorphic types and higher-order functions [3], approximating the features of mainstream programming languages like Java or C♯. This extension enables users to specify the same restrictions that the final program in a mainstream language will have. A bidirectional tree generation algorithm that supports polymorphic types has been shown to reduce the search space exponentially [17]. Our idea in this Chapter is to further advance the idea of using advanced type systems to encode semantic information in the tree structure, which was introduced as a more usable alternative to Grammar-Guided GP [7].

3.3.2 Grammar-Guided GP (GGGP)

Grammar-Guided GP generalized STGP by allowing the user to define the grammar of the program language, instead of just restricting the combination of symbols using types. In fact, the monomorphic STGP can easily be implemented using a Context-Free Grammar (CFG). The additional expressive power of grammars allowed GP to be applied to more SBSE domains, like Software Testing [37], Algorithmic Design [23] and Program Synthesis [8]. Grammar-based approaches typically use CFGs because the initialization algorithm is guaranteed to terminate in well-formed grammars (and a feasible maximum depth). As an example, take the CFG described in Backus Normal Form (BNF).

The algorithm for generating a random program is direct from its structure: starting in $\langle s \rangle$, recursively replace all non-terminals with one random choice from the possible alternatives. This (mostly useless) grammar only has an alternative in the root non-terminal, which supports just the two following programs: `k1 k2 k3 f` and `k2 k3 f`. So the only restriction that can prevent a program from being generated is the maximum allowed depth. If it is set to 2, there is no program that can be generated with just two recursion levels. Some GGGP approaches would discard the individual, while others would leave it in the population with a low fitness value, but both solutions have the drawback of spending time recursively generating tree components that are going to be discarded or ignored. This issue is more critical in recursive grammars (consider an extra alternative $\langle a \rangle ::= x \langle s \rangle$ was added to the grammar), because the proportion of programs that would reach the maximum allowed depth increases with the branching factor of non-terminals. In practice, for simple grammars that have a base case, this issue is solved by ignoring recursive productions when close to the maximum allowed depth. In Sect. 3.6, we will revisit this issue in a more complex scenario.

While CFGs present an efficient initialization procedure, they are very restricted in their expressiveness. It is simply not possible to model popular programming languages like Java or Rust using a CFG. One simple example is the definition of new variables, as the use of that variable is restricted to statements after the definition and in the same scope. More complex features like Rust lifelines or Contracts are also impossible to model using CFGs.

Christiansen Grammar Evolution [26] (CGE) was proposed to encode semantic properties of trees in the grammar. Christiansen Grammars are also called Adaptive Grammars because the productions can be modified as the grammar is recursively explored. This approach is able to solve the issue of variable declarations in programs, because the grammar can be expanded with the new variable after the declaration statement is visited. Despite the much higher expressive power, there are two major drawbacks with this approach, the first one being that it is computationally more intensive. Because it is necessary to keep track of changes to the grammar while producing programs, less programs can be generated in the same amount of time. So this is a crucial trade-off between how much the adaptive grammar is restricting the search space and the computational effort of keeping track of grammatical changes. The second drawback has to do with usability. There is a reason why programming language compilers have a simple grammar, and a second phase for detecting and discarding semantically invalid programs. Type systems are simple to implement on their own by transversing the tree, instead of encoding several rules as meta grammar rules.

3.3.3 Refined-Typed GP (RTGP)

Refined-Typed GP [7] was proposed to address this usability issue. It follows the line of work of STGP by having a simple grammar and using an advanced type system with dependent types to have the same expressive power as Christiansen Grammars. For the user, there is a substantial difference: they do not have to change the grammar (as a very simple lambda calculus-inspired language is used), but instead describe what variables and functions are available in the context, and use refinements to specify the semantic restrictions. Besides the usability aspect, another advantage of RTGP over CGE is the fact that a subset of the restriction language can be converted into Verification Conditions that are statically deemed valid or not through the use of an SMT solver. This is the basis of Liquid Types [31], which have been used to address several SBSE issues like Program Synthesis [13], resource analysis [16], communication and synchronization verification [12, 15], information flow [29] and distributed data replication [18]. Furthermore, there are no practical approaches for these problems that rely on Christiansen Grammars.

3.4 Correct-by-Construction Versus Generate-and-Validate

CGE follows a correct-by-construction approach: any (maximum depth-permitting) generated tree is valid according to the meta-rules encoded in the grammar. RTGP is more flexible: synthesis can be type-aware, thus correct-by-construction, or simple types (corresponding to CFG with a minimal context) can be used to generate trees, which are then validated according to the full refined types, thus following a generate-and-validate procedure. As an example, imagine we want to generate programs with the type `(x:Int | x > 3) -> s:List[String] | len(s) == x@` (a function that receives an integer that is guaranteed to be greater than 3, and returns a string of size equal to the input integer) and the replicate function, that takes an element and an integer and returns a list with as many elements as the integer value, of type (e:A, k:Int) → l:List[A] | len(l) == k in the context. In this case, a type-aware synthesis could take the following steps among the many allowed by the grammar (?n : T denotes the nth yet-to-be-synthesized hole of type T):

```
1  ?1 : (x:Int | x > 3) → {s:List[String] | len(s) == x}
2  \x → (?2 : {s:List[String] | len(s) == x })
3  \x → replicate(?3:String, ?4:Int)
4  \x → replicate("abc123", x)
```

The advantages of RTGP can be seen in the last step. The replicate function can only synthesize x as its second argument in a type-safe manner (because the replicate type guarantees that the size of the output is equal to the second argument, and we know that the size of the output list is equal to the variable x). A generate-and-validate approach would synthesize any function that generates lists of strings, including, for

$$
\begin{aligned}
\langle t \rangle ::= &\ \mathrm{l} \\
| &\ \mathrm{x} \\
| &\ \langle t \rangle\ \langle t \rangle \\
| &\ \lambda \mathrm{x} \mapsto \langle t \rangle \\
| &\ \text{if } \langle t \rangle \text{ then } \langle t \rangle \text{ else } \langle t \rangle
\end{aligned}
\qquad
\begin{aligned}
\langle T \rangle ::= &\ \mathrm{B} \\
| &\ (\mathrm{x}{:}\langle T \rangle) \to \mathrm{T} \\
| &\ \{\mathrm{x}{:}\langle T \rangle \mid \mathrm{r}\ \}
\end{aligned}
$$

Fig. 3.1 The ÆON grammar for liquid types

instance, replicate("a", 2), which would later be discarded or ignored. Synthesizing candidate programs that will be discarded serves no purpose in the search and should normally be avoided.

To understand the exceptions, take this very simple example: {x:Int | x != 42}. A type-aware approach would be much slower than generating a simple integer and verifying that it is not equal to that particular value. Given the full range of integer, the chance of synthesizing 42 is so close to zero, that a simple call to an SMT solver would be orders of magnitude slower, even in the case where 42 is generated the first time and a second number is required. On the other end of the spectrum, to synthesize {x:Int | x == 43} the tiny probability of missing the target type of the previous example is now the tiny probability of finding the right value. The overhead of calling the SMT solver for this simple example would be negligible compared to the average case of a random search. Of course, in real-world examples most expressions come from more complex types that are not so black and white.

For the remainder of this Chapter we will address the issue of implementing type-aware synthesis, as it is the most challenging scenario. Existing STGP and GGGP already address the implementation of generate-and-validate programs by first generating the programs, and only then validating whether that program fulfills the specification, discarding or ignoring it.

Let us now understand the synthesis procedure used in the previous example. We are using the ÆON programming language [7], which is inspired by ML and its Liquid version [31]. ÆON has a very simple core grammar (Fig. 3.1) of terms (t) and types (T). We use the following meta-variables: `l` for boolean and integer literals; `B` for base types like `Int` or `Bool`; `x` for variable names and `r` for valid refinements.

Liquid type-checking and inference is implemented with the same rules of Sprite [14] and term synthesis is symmetrical to type-checking, following the same approach of Synquid [28]. These rules will be explained as they are needed throughout the document.

Having a mental model of the grammar of this language in mind, we can now understand two major performance issues of type-aware synthesis in the context of GGGP. We consider two individual representations: the tree-representation of the original GGGP and the Gramatical Evolution [32] (GE) representation that uses variable-sized lists of integers.

The first issue arrises when an unproductive rule is chosen non-stop. Take the example of synthesizing a term of type {x:Int | x > y} and y:Int is in the context. If either a random number generator chooses the first of the five rules or the GE codon is an integer multiple of 5, the algorithm will try to synthesize an integer literal. However, the SMT solver will output that there is no literal that can always be greater than y for any given y (y can always take the value of the maximum supported integer). In that case the algorithm will backtrack to the previous recursion level and try another path based on random choice or the next codon. If the recursion is at the maximum allowed depth, then the probability of choosing an unproductive rule is 50% (choice between literal and variable). While this issue seems only theoretical in the sense that repetitions in random choices seem improbable, our experience implementing ÆON shows that this is a common problem that substantially impacts performance. The main reason is that this unproductive behavior can occur at every level of the term synthesis. Because of the binary and ternary branching present in the grammar, population initialization slows down as the maximum allowed depth increases. Section 3.7 will present evidence of this issue in practice.

The second issue is shared with traditional GGGP: when the maximum depth is reached without a valid terminal symbol (there is no literal nor variable that type-checks), the standard approach is to either discard or assign a low fitness value (which will, in practice, ignore the individual during the next selection phase). While this may be acceptable in CFG grammars, the expensive cost of type-aware synthesis makes discarding or ignoring individuals much more detrimental to the overall search efficiency.

3.5 Direct Versus Indirect Representations

In our Program Synthesis tool for the ÆON language, we have experimented with a tree representation. We have found that crossover and mutation operators on tree representations had a negative impact on the performance of the algorithm. To perform either operation, the tree would need to be transversed again to rebuild the necessary context at the crossover/mutation point. Caching the context inside the tree did not improve the process, as allocating memory also takes time (not having cache promotes memory reusing via garbage collection).

We have also considered using a multiple-stack representation inspired by the PushGP system [36], but it is not feasible because it is not possible to group values in stacks of different types, since there are infinite types. Even if only the types used in a program have their own dedicated stack, values can be in many different types and each stack may have only a single value, breaking the constraints of the system that provide an efficient implementation.

Structured [20] and Dynamically Structured Grammatical Evolution [19] (SGE and DSGE) could have been implemented, partitioning the list of integers in sublists for each of the grammar rules. This approach improves the locality of the genotype-phenotype mapping, but in the context of our grammar, it provides limited gain because there are only two non-terminals (terms and types), and types occur only on a single term production (the application rules).

We settled on a representation based on a variable-length list of integers. The crossover and mutation operators are standard in Grammatical Evolution [32] (GE), with which we share the representation. The main innovation of this approach is the mapping function between the genotype (list of integers) and the phenotype (program trees that are guaranteed to be valid with regard to a logical specification, i.e., programs that type-check).

3.6 A Dynamic Grammar-Guided Mapping

An ideal mapping function would promote diversity [2], have a good locality [20], have a perfect success rate and be fast. We focus on the last two goals, as without them, it is impossible to even evaluate the first two. Total time consumption of individual synthesis is important to be able to use this approach in real-world problems in reasonable time. Having a low success rate will contribute to an higher execution total, because in order to generate a population of 100 individuals having a 50% success rate, one would need to wait the time of generating 200 individuals.

3.6.1 *GE Mapping*

Following the spirit of GE, we will deterministically generate a valid program tree from a list of integers, or fail if that task is impossible within given restrictions (maximum allowed depth and backtracking budget). Like in GE, we keep track of the current index of the array that will help make the next decision (based on grammar non-determinism). If we have the following two productions that can expand the current rule (S): $S ::= AB$ and $S ::= c$, that choice will be made using the value at the current index (v). Because there are only two choices, we will take the first if the value is even or the second if the value is odd ($choice = v\%alts$, where $alts$ is the number of alternatives). After making this choice, we move the current index to the following position of the array. If the index overflows the size of the array, it wraps around to the beginning of the array.

3.6.2 *Semantic Filter of Valid Productions*

The first difference from GE is that not all valid productions are considered when taking the decision at each step. GE typically only excludes rules that will expand the program beyond the maximum allowed depth. In ÆON, if the current allowed depth is 0, only variables and literals will be considered.

Our approach adds another filter layer that is semantic. Before considering a production, there is a check to validate that the production will be able to produce

```
1  def term_synthesis(ctx, type, d):
2      options = []
3      if ctx.has_lit_of_type(type):
4          options.add(lit)
5      if ctx.has_var_of_type(type):
6          options.add(var)
7      if d > 0:
8          options.add(app)
9          if type.is_abstraction():
10             options.add(abs)
11         options.add(ifthenelse)
12
13     while options and has_budget():
14         checkpoint()
15         try:
16             return follow_rule(choice(options))
17         except NoBudget:
18             rollback()
19     raise NoBudget()
```

Fig. 3.2 High-level synthesis algorithm with the semantic filter. The function term_synthesis is responsible for synthesizing a term based on the context (`ctx`), the expected type (`type`) and how many depth levels are left (`d`). The `lit`, `var`, `app`, `abs` and `ifthenelse` correspond to the recursive functions that implement each of the expansions of the ÆON grammar in Fig. 3.1. The `checkpoint` and `rollback` functions support the dynamic probability manager

a result. The variable production will only be available for choosing if there is at least one variable in the program context that is of the required type (e.g., if there are no variables of type {x:Int | x == 3}, only literals will be considered). If, on the other hand, there is no valid literal for that type, the rule will not be considered (e.g., {x:Int | x == y}, with `y` of type `Int` in the context).

This filter is important to be applied as early in the process as possible. To understand this, let us consider the array [1, 11, 33] (as probable to occur as any other of the same size). Without the filter, we would choose the variable production (first index is odd), figure out that there are no variables of the target type and backtrack, advancing to the second index. This would occur also on the following tries because all the next values in the genotype are odd. And because we wrap around when we get to the end of the array, this process would continue for ever (or, in practice, until we reach the maximum backtracking budget) without making any progress. Because the ÆON grammar is so simple, this issue occurs very frequently and would make this approach unusable without the semantic filter.

Figure 3.2 depicts the necessary conditions to implement the context-dependent semantic filter in the ÆON language. The `checkpoint` and `rollback` functions will be explained later, as they serve another purpose.

3.6.3 Dynamic and Depth-Aware Dynamic Approaches

Even with the semantic filter, there are several common cases in which the algorithm reaches a point where it makes no progress. And, unlike in the previous examples, it is not possible to identify those cases.

Let us consider the example of synthesizing a term of type (x:A)→ Bool, with a maximum allowed depth of 2, and no variables in the program context. Because there are no variables of this type, and only possible recursions in the grammar expansion, the only valid solutions are (x→ true and x→ false), corresponding to selecting the abstraction (`abs`) production (depth = 1), then the literal (`lit`) production with one of either true or false (depth=0). However, while still having depth 1 (meaning that we only recurse once) all `abs`, `app` and `ifthenelse` productions are valid, even after the semantic filter, which excluded the `lit` and `var` productions). If the `ifthenelse` is selected, the condition will be generated and when building the *then* term it will fail to find an alternative with depth 0 (as we already knew there was no literal or variable of that type). This example evidences this problem near the leafs, but in the case we had a larger maximum allowed depth, and the if rule was chosen n recursion levels before the leafs, we would experience the same issue. Because of the low number of productions, this issue also occurs frequently in practice.

To improve this scenario, we propose a dynamic approach to population initialization. Similar approaches have been taken by Criado et al. [30], to create a uniform population initialization algorithm, and by Mégane et al. [21], that adapts probabilities of expansions during the evolution process. However, none of the approaches is feasible in our scenario because they are designed for CFGs, and while our grammar is context-free, the semantic validation is not context-free.

Figure 3.3 provides an illustrative example of how the probabilities are dynamically adjusted. The overall algorithm relies on an initial probability[1] for all rules (100 in the example). This value will be reduced during the population initialization phase. Each time a term is synthesized (even recursively), the decision of which node to use will be made from the value at the current index in the genotype array. Instead of using the modulo operation to select one of the options like in GE, this operation takes into consideration the probabilities of each rule, similarly to how the Roulette Wheel operator [27] is implemented. The synthesis process continues recursively to each subterm. If the tree is generated, no changes are made to the probabilities. However, if the backtracking budget is exceeded, we record each decision made in the relevant subterms, and we decrease by 10% the probability of that rule. The algorithm in Fig. 3.2 includes the `checkpoint` and `rollback`, which are used to implement the tracking of decisions made at each level. This is important to only penalize decisions made within this term, and not other decisions made higher-up in the tree.

Note that, because rule probabilities are dynamic, the phenotypical mapping of the genotype of the first individual might not be the same after the population initialization process, because weights have changed. To overcome this, the genotype

[1] We use the term 'probability' loosely, not with a statistical meaning.

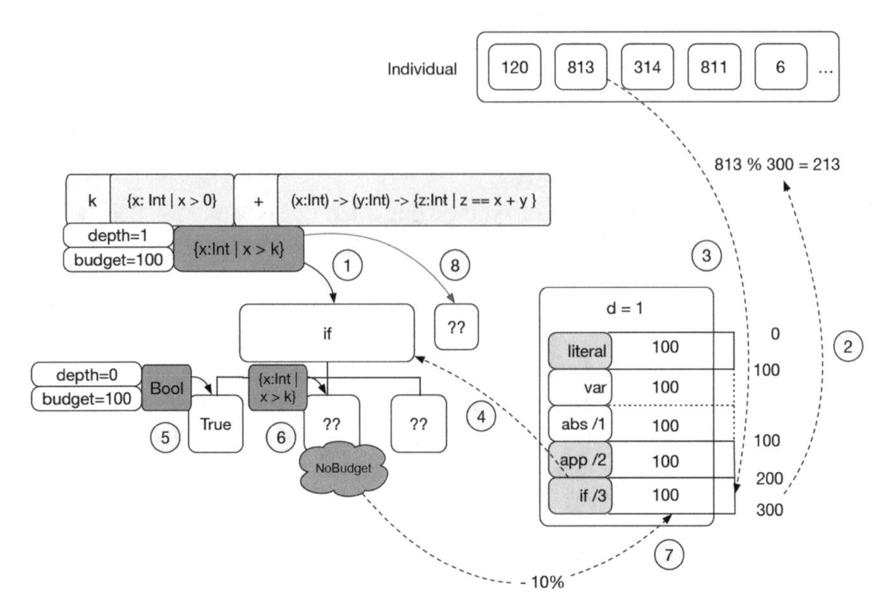

Fig. 3.3 A backtracking situation. (1) In the context described in yellow, at depth 1 and budget 100, the system will synthesize an expression of type Integer greater than k; (2) The total sum of **valid** (in green) rules is computed; (3) The index for the selected rule is computed from the current value at the genotype modulo the total value computed; (4) The rule corresponding to that index is chosen, and the necessary subtrees are synthesized; (5) At depth 0 and without boolean variables, the literal rule is chosen automatically, and using the 314 the True value is chosen; (6) At depth 0, there are no possible literals or variables greater than an unknown k, thus a backtracking event occurs; (7) Any rule chosen in this inner synthesis has its probability decreased by 10%; (8) The synthesis for the original type is retried, with a budget of 99

representation of individuals can be reverse engineered from the final weights in linear time to the number of nodes, since instead of the original genotype, only the recorded decisions need to be saved. This overhead is relatively small compared to the exponential cost of unnecessary backtracking.

The goal of this adjustment is to prevent future similar dead-end scenarios (that exhaust backtracking) to occur. Because the problem is context-dependent, the actual probability should also be context-dependent. However, contexts are very volatile and diverse (each abstraction introduces a new variable and type in the context), which means they are not shared across individuals, thus not very helpful.

To overcome this limitation, we also propose a depth-aware dynamic approach that uses the current depth as a proxy for the context that has the advantage of being shared across all individuals, despite not having a one-to-one correspondence with the context.

3.7 Evaluation

We evaluate the three variants described above (Semantic Filter, Dynamic and Depth-Aware Dynamic) and compare them with the baseline (GE) within the ÆON programming language on a AMD Ryzen Threadripper 3960X (24 cores) with 64GB of RAM.

Our experiments consist on different runs trying to synthesize 100 terms of type {x:Int | x > 0 && x < 1000}. The initial program context contains the standard mathematical library of ÆON (logical and arithmetic operators). The backtracking budget is fixed at 100 for all individuals. Each synthesis run is repeated 30 times.

All of the approaches convert the genotype of 100 individuals to the corresponding phenotype. The genotypes are list of random integers of length between 10 and 100. Because our grammar-guided approach uses a modulo-based mapping, the size of each individual matters only for the diversity of available numbers in the genotype.

We report results for this very simple type for ease of read and because the probability of finding a value using generate-and-validate approaches is really low ($100/2^{64} = 5.42^{-18}$) and represents the cases where correct-by-construction approaches should be favored. Traditionally, this example can be easily solved by generating a random integer between 1 and 1000, but we are using this type as a readable proxy for equivalent expressions such as {x:Int | (x >= 21 && x < y)|| x < (2*10)|| x == (80/z)}, where y and z are in context with types {y:Int | 10 * y == 10000} and {z:Int| 2+z == 6 }, whose bounds would not be obtained faster than using an SMT solver.

Figures 3.4, 3.5, 3.6 and 3.7 depict the percentage of successes (measured as the number of trees generated in 100 tries), the average number of successes per second, the diversity among the population (measured using the tree distance [40]) and the average depth of the population, all for executions with maximum tree depths between 5 and 13.

Figure 3.4 shows that, as the maximum tree depth increases, the probability of finding viable candidates decreases. This behaviour is to be expected, as having more leaf nodes increases the likelihood of reaching dead-ends. This occurs because the backtracking budget is fixed (representing the time constraint). The plain (depth-independent) Dynamic probability management is the one reaching more successes, and Fig. 3.5 confirms that this is indeed the most productive approach. The Depth-Aware Dynamic approach is the one that generates more diverse trees, both in terms of tree distance (Fig. 3.6) and average depth (Fig. 3.7), as it reduces the bias of always choosing the same rules. However, it does so at the cost of having less successes per second (Fig. 3.5). Figures 3.6 and 3.7 also evidence that all proposed approaches outperform GE with regard to both types of diversity in the generated population. While our approach does not generate the exact depth requested by the user, it is able to generate values closer to the desired depth than GE. In fact, this variance naturally mimics the effect of the grow initialization method. As expected, the Semantic Filter approach is not as good as the dynamic approaches. Finally, we notice that the three variants behave more similarly to each other when maximum tree depth is lower.

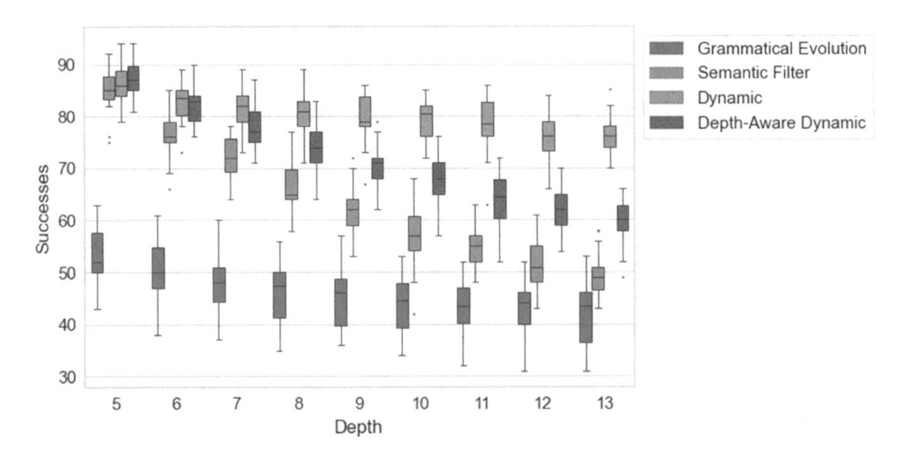

Fig. 3.4 Number of trees generated in 100 tries. Higher is better

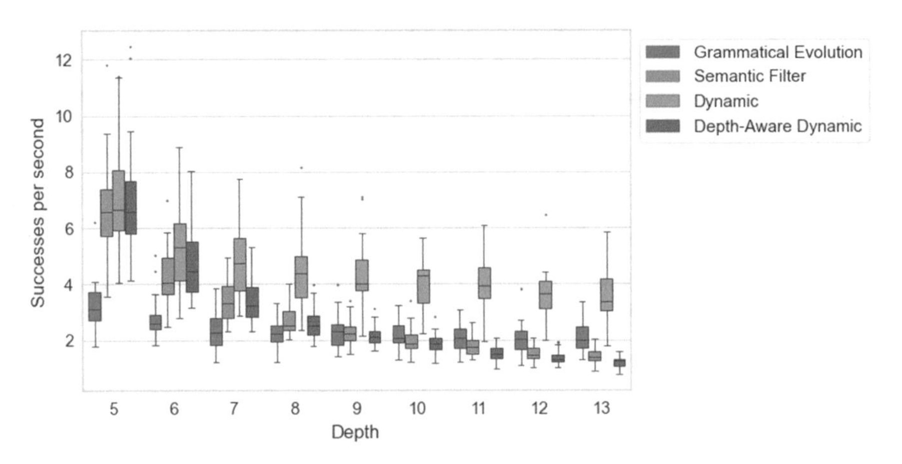

Fig. 3.5 Number of successes per second. Higher is better

3.8 Conclusions

We have identified the need for GP methods to encode program semantics not only as a validation step, but as constraints on the creation of individuals, especially in the field of SBSE. SMT and SAT solvers have been a useful tool in the different domains to perform efficient search on a limited formula language, thus it is desirable to incorporate those tools inside the evolutionary process of GP.

We discussed the possible implementations for the integration of Liquid Types into GP, concluding that a grammar-guided approach with a type-safe mapping is a feasible approach. We presented an initialization algorithm that dynamically adjusts the probabilities of grammar productions according to its failures in expanding the grammar up to a given depth. We showed that our approach is beneficial, almost

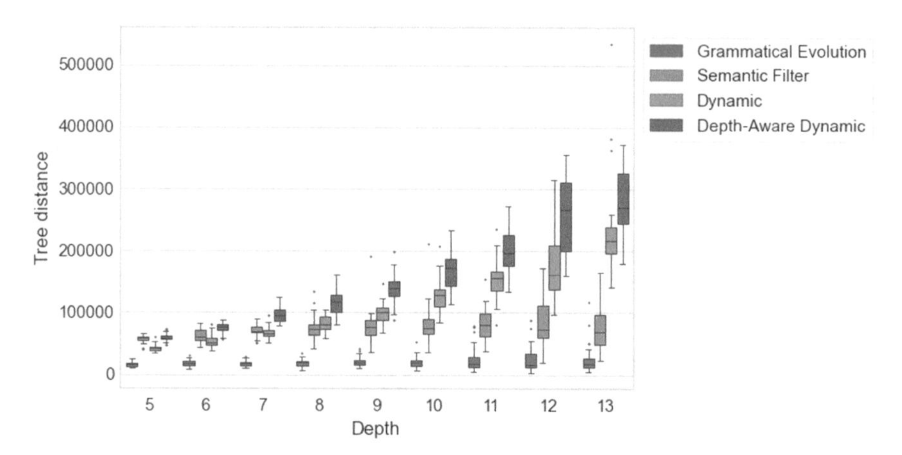

Fig. 3.6 Level of diversity in the generated individuals, measured by tree distance. Higher is better

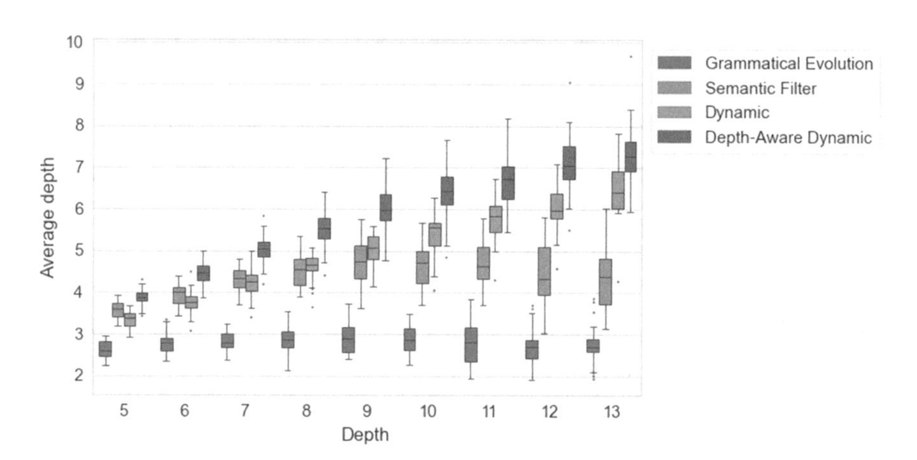

Fig. 3.7 Average depth of the generated individuals. A combination of higher values and wider range is better

regardless of the variant used, when generating values between very tight bounds, which are exactly the cases that require a more efficient initialization, do the rate of failed validations. Furthermore, we showed that our three variants have better performance than Grammatical Evolution, and generate a more diverse population. We also identified a trade-off between generating more individuals by using a depth-independent approach or having individuals with higher depth and diversity by using a depth-aware approach.

For future work, we intend to explore how to efficiently implement other operators (like mutation and crossover) in the same context, and benchmark RTGP against GGGP.

Acknowledgements This work was supported by the Fundação para a Ciência e a Tecnologia (FCT) under LASIGE Research Unit (UIDB/00408/2020 and UIDP/00408/2020), the GADgET project (DSAIPA/DS/0022/2018) and the CMU Portugal project CAMELOT (POCI-01-0247-FEDER-045915).

References

1. Alhroob, A., Imam, A.T., Al-Heisa, R.: The use of artificial neural networks for extracting actions and actors from requirements document. Inf. Softw. Technol. **101**, 1–15 (2018)
2. Bartoli, A., Lorenzo, A.D., Medvet, E., Squillero, G.: Multi-level diversity promotion strategies for grammar-guided genetic programming. Appl. Soft Comput. **83** (2019)
3. Binard, F., Felty, A.P.: Genetic programming with polymorphic types and higher-order functions. In: Ryan, C., Keijzer, M. (eds.) Genetic and Evolutionary Computation Conference, GECCO 2008, Proceedings, Atlanta, GA, USA, 12–16 July 2008, pp. 1187–1194. ACM (2008)
4. Bladek, I., Krawiec, K.: Evolutionary program sketching. In: McDermott, J., Castelli, M., Sekanina, L., Haasdijk, E., García-Sánchez, P. (eds.) Genetic Programming—20th European Conference, EuroGP 2017, Amsterdam, The Netherlands, 19–21 April 2017, Proceedings, Lecture Notes in Computer Science, vol. 10196, pp. 3–18 (2017)
5. Bojarczuk, C.C., Lopes, H.S., Freitas, A.A., Michalkiewicz, E.L.: A constrained-syntax genetic programming system for discovering classification rules: application to medical data sets. Artif. Intell. Med. **30**(1), 27–48 (2004)
6. Durieux, T., Madeiral, F., Martinez, M., Abreu, R.: Empirical review of java program repair tools: a large-scale experiment on 2, 141 bugs and 23, 551 repair attempts. In: Dumas, M., Pfahl, D., Apel, S., Russo, A. (eds.) Proceedings of the ACM Joint Meeting on European Software Engineering Conference and Symposium on the Foundations of Software Engineering, ESEC/SIGSOFT FSE 2019, Tallinn, Estonia, 26–30 Aug 2019, pp. 302–313. ACM (2019)
7. Fonseca, A., Santos, P., Silva, S.: The usability argument for refinement typed genetic programming. In: Bäck, T., Preuss, M., Deutz, A.H., Wang, H., Doerr, C., Emmerich, M.T.M., Trautmann, H. (eds.) Parallel Problem Solving from Nature—PPSN XVI—16th International Conference, PPSN 2020, Leiden, The Netherlands, 5–9 Sept 2020, Proceedings, Part II, Lecture Notes in Computer Science, vol. 12270, pp. 18–32. Springer (2020)
8. Forstenlechner, S., Fagan, D., Nicolau, M., O'Neill, M.: Extending program synthesis grammars for grammar-guided genetic programming. In: Auger, A., Fonseca, C.M., Lourenço, N., Machado, P., Paquete, L., Whitley, L.D. (eds.) Parallel Problem Solving from Nature—PPSN XV—15th International Conference, Coimbra, Portugal, 8–12 Sept 2018, Proceedings, Part I, Lecture Notes in Computer Science, vol. 11101, pp. 197–208. Springer (2018)
9. Fraser, G., Arcuri, A.: Evosuite: automatic test suite generation for object-oriented software. In: Gyimóthy, T., Zeller, A. (eds.) SIGSOFT/FSE'11 19th ACM SIGSOFT Symposium on the Foundations of Software Engineering (FSE-19) and ESEC'11: 13th European Software Engineering Conference (ESEC-13), Szeged, Hungary, 5–9 Sept 2011, pp. 416–419. ACM (2011)
10. Galeotti, J.P., Fraser, G., Arcuri, A.: Extending a search-based test generator with adaptive dynamic symbolic execution. In: Pasareanu, C.S., Marinov, D. (eds.) International Symposium on Software Testing and Analysis, ISSTA '14, San Jose, CA, USA—21–24 July 2014, pp. 421–424. ACM (2014)
11. Goues, C.L., Nguyen, T., Forrest, S., Weimer, W.: Genprog: a generic method for automatic software repair. IEEE Trans. Softw. Eng. **38**(1), 54–72 (2012)
12. Griffith, D., Gunter, E.L.: Liquidpi: inferrable dependent session types. In: Brat, G., Rungta, N., Venet, A. (eds.) NASA Formal Methods, 5th International Symposium, NFM 2013, Moffett Field, CA, USA, 14–16 May 2013. Proceedings, Lecture Notes in Computer Science, vol. 7871, pp. 185–197. Springer (2013)

13. Guo, Z., James, M., Justo, D., Zhou, J., Wang, Z., Jhala, R., Polikarpova, N.: Program synthesis by type-guided abstraction refinement. Proc. ACM Program. Lang. **4**(POPL), 12:1–12:28 (2020)
14. Jhala, R., Vazou, N.: Refinement types: a tutorial. CoRR (2020). https://arxiv.org/abs/2010.07763
15. Kloos, J., Majumdar, R., Vafeiadis, V.: Asynchronous liquid separation types. In: Boyland, J.T. (ed.) 29th European Conference on Object-Oriented Programming, ECOOP 2015, 5–10 July 2015, Prague, Czech Republic, LIPIcs, vol. 37, pp. 396–420. Schloss Dagstuhl—Leibniz-Zentrum für Informatik (2015)
16. Knoth, T., Wang, D., Reynolds, A., Hoffmann, J., Polikarpova, N.: Liquid resource types. Proc. ACM Program. Lang. **4**(ICFP), 106:1–106:29 (2020)
17. Kren, T., Moudrík, J., Neruda, R.: Combining top-down and bottom-up approaches for automated discovery of typed programs. In: 2017 IEEE Symposium Series on Computational Intelligence, SSCI 2017, Honolulu, HI, USA, Nov 27–Dec 1, 2017, pp. 1–8. IEEE (2017)
18. Liu, Y., Parker, J., Redmond, P., Kuper, L., Hicks, M., Vazou, N.: Verifying replicated data types with typeclass refinements in liquid haskell. Proc. ACM Program. Lang. **4**(OOPSLA), 216:1–216:30 (2020)
19. Lourenço, N., Assunção, F., Pereira, F.B., Costa, E., Machado, P.: Structured grammatical evolution: a dynamic approach. In: Ryan, C., O'Neill, M., Collins, J.J. (eds.) Handbook of Grammatical Evolution, pp. 137–161. Springer (2018)
20. Lourenço, N., Pereira, F.B., Costa, E.: SGE: A structured representation for grammatical evolution. In: Bonnevay, S., Legrand, P., Monmarché, N., Lutton, E., Schoenauer, M. (eds.) Artificial Evolution—12th International Conference, Evolution Artificielle, EA 2015, Lyon, France, 26–28 Oct 2015. Revised Selected Papers, Lecture Notes in Computer Science, vol. 9554, pp. 136–148. Springer (2015)
21. Mégane, J., Lourenço, N., Machado, P.: Probabilistic grammatical evolution. In: Hu, T., Lourenço, N., Medvet, E. (eds.) Genetic Programming—24th European Conference, EuroGP 2021, Held as Part of EvoStar 2021, Virtual Event, 7–9 April 2021, Proceedings, Lecture Notes in Computer Science, vol. 12691, pp. 198–213. Springer (2021)
22. Meudec, C.: ATGen: automatic test data generation using constraint logic programming and symbolic execution. Softw. Test. Verif. Reliab. **11**(2), 81–96 (2001)
23. de Miranda, P.B.C., Prudêncio, R.B.C.: Generation of particle swarm optimization algorithms: an experimental study using grammar-guided genetic programming. Appl. Soft Comput. **60**, 281–296 (2017)
24. Montana, D.J.: Strongly typed genetic programming. Evol. Comput. **3**(2), 199–230 (1995)
25. Nguyen, H.D.T., Qi, D., Roychoudhury, A., Chandra, S.: SemFix: program repair via semantic analysis. In: Notkin, D., Cheng, B.H.C., Pohl, K. (eds.) 35th International Conference on Software Engineering, ICSE '13, San Francisco, CA, USA, 18–26 May 2013, pp. 772–781. IEEE Computer Society (2013)
26. Ortega, A., de la Cruz, M., Alfonseca, M.: Christiansen grammar evolution: grammatical evolution with semantics. IEEE Trans. Evol. Comput. **11**(1), 77–90 (2007)
27. Poli, R., Langdon, W.B., McPhee, N.F.: A Field Guide to Genetic Programming. Lulu Enterprises, UK Ltd. (2008)
28. Polikarpova, N., Solar-Lezama, A.: Program synthesis from polymorphic refinement types. CoRR (2015). http://arxiv.org/abs/1510.08419
29. Polikarpova, N., Stefan, D., Yang, J., Itzhaky, S., Hance, T., Solar-Lezama, A.: Liquid information flow control. Proc. ACM Program. Lang. **4**(ICFP), 105:1–105:30 (2020)
30. Ramos-Criado, P., Rolanía, D.B., Manrique, D., Serrano, E.: Grammatically uniform population initialization for grammar-guided genetic programming. Soft Comput. **24**(15), 11265–11282 (2020)
31. Rondon, P.M., Kawaguchi, M., Jhala, R.: Liquid types. In: Gupta, R., Amarasinghe, S.P. (eds.) Proceedings of the ACM SIGPLAN 2008 Conference on Programming Language Design and Implementation, Tucson, AZ, USA, 7–13 June 2008, pp. 159–169. ACM (2008)

32. Ryan, C., Collins, J.J., O'Neill, M.: Grammatical evolution: evolving programs for an arbitrary language. In: Banzhaf, W., Poli, R., Schoenauer, M., Fogarty, T.C. (eds.) Genetic Programming, First European Workshop, EuroGP'98, Paris, France, 14–15 April 1998, Proceedings, Lecture Notes in Computer Science, vol. 1391, pp. 83–96. Springer (1998)
33. Santos, P., Campos, J., Timperley, C.S., Fonseca, A.: Augmenting search-based techniques with static synthesis-based input generation. In: ICSE '21: 42nd International Conference on Software Engineering, Workshops, Madrid, Spain, 17 May—4 June 2021. ACM (2021)
34. Sato, Y.: Specification-based test case generation with constrained genetic programming. In: 20th IEEE International Conference on Software Quality, Reliability and Security Companion, QRS Companion 2020, Macau, China, 11–14 Dec 2020, pp. 98–103. IEEE (2020)
35. Solar-Lezama, A.: The sketching approach to program synthesis. In: Hu, Z. (ed.) Programming Languages and Systems, 7th Asian Symposium, APLAS 2009, Seoul, Korea, 14–16 Dec 2009. Proceedings, Lecture Notes in Computer Science, vol. 5904, pp. 4–13. Springer (2009)
36. Spector, L., Robinson, A.J.: Genetic programming and autoconstructive evolution with the push programming language. Genet. Program. Evolvable Mach. **3**(1), 7–40 (2002)
37. Vergilio, S.R., Pozo, A.T.R.: A grammar-guided genetic programming framework configured for data mining and software testing. Int. J. Softw. Eng. Knowl. Eng. **16**(2), 245–268 (2006)
38. Wappler, S., Wegener, J.: Evolutionary unit testing of object-oriented software using strongly-typed genetic programming. In: Cattolico M. (ed.) Genetic and Evolutionary Computation Conference, GECCO 2006, Proceedings, Seattle, Washington, USA, 8–12 July 2006, pp. 1925–1932. ACM (2006)
39. Yuan, Y., Banzhaf, W.: ARJA: automated repair of java programs via multi-objective genetic programming. IEEE Trans. Softw. Eng. **46**(10), 1040–1067 (2020)
40. Zhang, K., Shasha, D.E.: Simple fast algorithms for the editing distance between trees and related problems. SIAM J. Comput. **18**(6), 1245–1262 (1989)

Chapter 4
What Can Phylogenetic Metrics Tell us About Useful Diversity in Evolutionary Algorithms?

Jose Guadalupe Hernandez, Alexander Lalejini, and Emily Dolson

Abstract It is generally accepted that "diversity" is associated with success in evolutionary algorithms. However, diversity is a broad concept that can be measured and defined in a multitude of ways. To date, most evolutionary computation research has measured diversity using the richness and/or evenness of a particular genotypic or phenotypic property. While these metrics are informative, we hypothesize that other diversity metrics are more strongly predictive of success. Phylogenetic diversity metrics are a class of metrics popularly used in biology, which take into account the evolutionary history of a population. Here, we investigate the extent to which (1) these metrics provide different information than those traditionally used in evolutionary computation, and (2) these metrics better predict the long-term success of a run of evolutionary computation. We find that, in most cases, phylogenetic metrics behave meaningfully differently from other diversity metrics. Moreover, our results suggest that phylogenetic diversity is indeed a better predictor of success.

4.1 Introduction

Maintaining a sufficiently diverse population to successfully solve challenging problems is a central challenge in all branches of evolutionary computation. If the population's diversity collapses, an evolutionary algorithm can prematurely converge on a sub-optimal solution from which it is unable to escape [10]. While many diversity

J. G. Hernandez · E. Dolson (✉)
BEACON Center for the Study of Evolution in Action and Department of Computer Science and Ecology, Evolutionary Biology, and Behavior Program, Michigan State University, East Lansing, MI, USA
e-mail: dolsonem@msu.edu

J. G. Hernandez
e-mail: herna383@msu.edu

A. Lalejini
Department of Ecology and Evolutionary Biology, University of Michigan, Ann Arbor, MI, USA
e-mail: amlalejini@gmail.com

W. Banzhaf et al. (eds.), *Genetic Programming Theory and Practice XVIII*,
Genetic and Evolutionary Computation,
https://doi.org/10.1007/978-981-16-8113-4_4

maintenance techniques have been designed to combat this challenge, we currently lack a clear understanding of what factors contribute to their success or failure in any given situation. Broadly speaking, diversity maintenance techniques can fail in two ways: (1) failure to maintain a diverse population at all, and (2) failure to maintain diversity that is actually helpful to solving the problem. Because more effort has historically been paid to the former category, here we will focus on the latter.

In evolutionary computation, diversity is usually evaluated by counting the number of unique "types" in the population. These "types" may be genotypes, phenotypes, ecotypes, species, or other descriptors of a group of solutions. The most commonly used types are phenotypes, which are often described using error vectors, behaviors, or output values. Sometimes a population's diversity is measured as the raw count of unique types (in biology, such metrics are called "richness"). Alternatively, sometimes different metrics are used that take into account how evenly the population is distributed across the types (as in Shannon diversity/entropy). Occasionally more nuanced metrics are used that consider the level of similarity of the types in the population (e.g., calculating Euclidean distance between error vectors, or including some sort of clustering step). However, the metrics commonly used in evolutionary computation are a small subset of the broader range of ways one could measure diversity. Prior work suggests that some diversity metrics are more predictive of success than others [17, 22], so investigating the implications of a wider variety of diversity metrics is worthwhile.

A variety of diversity metrics from ecology and evolutionary biology have yet to be examined in the context of evolutionary computation. One particularly interesting class of diversity metrics, called phylodiversity metrics [29], takes into account the evolutionary history of a population to calculate its diversity. For intuition behind how these metrics are different from other diversity metrics, see Fig. 4.1. They do so by measuring the topology of the phylogenetic tree for the population (i.e., the ancestry tree). Intuitively, types that are more evolutionarily distant from each other (i.e., share a more distant common ancestor) are likely farther apart from each other in the fitness landscape. As such, the ability to maintain evolutionarily distinct types may be important for effective problem-solving in evolutionary computing, due to the likely difficulty of re-evolving such distinct types from each other. Notably, biological simulations suggest that phylogenetic diversity provides different information about a population than a specific type of phenotypic richness (functional diversity) does; moreover, the extent of this difference varies across different scenarios [30].

Preliminary data supports the hypotheses that (1) phylogenetic diversity captures information that other diversity metrics do not [7, 30], and (2) phylogenetic diversity may be a better predictor of success in genetic programming than other diversity metrics [7]. Here, we investigate these hypotheses in the context of a range of problems and diversity maintenance techniques. Our findings provide further support for these two hypotheses. In particular, using a technique called causality analysis, we show that phylogenetic diversity is a stronger predictor of future fitness than phenotypic diversity is.

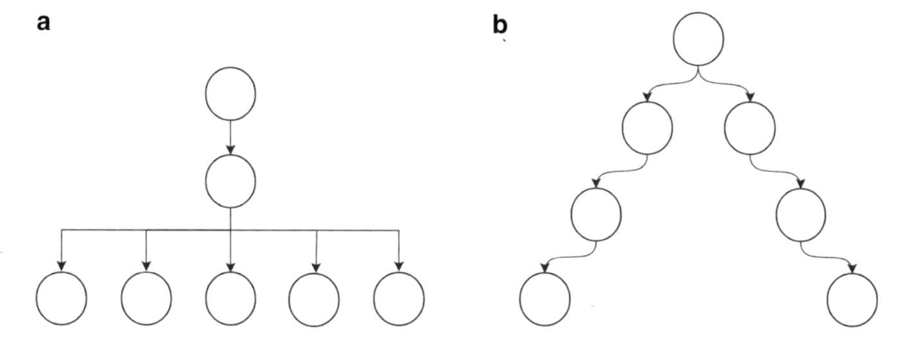

Fig. 4.1 Example populations with different kinds of diversity. This figure shows two different phylogenies. Arrows show parent-child relationships. Each node is a taxonomically unique phenotype (i.e., a phenotype with a unique evolutionary origin). For simplicity, leaf nodes in these diagrams are assumed to be the current set of taxa in the population; in reality, there could be non-leaf nodes corresponding to extant taxa. **a**. A population with high phenotypic diversity (phenotypic richness = 5) and low phylogenetic diversity (mean pairwise distance = 2). **b**. A population with low phenotypic diversity (phenotypic richness = 2) and high phylogenetic diversity (mean pairwise distance = 6)

4.2 Methods

To identify phylogenetic metrics that correlate with success in evolutionary computation, we ran an evolutionary algorithm on a variety of problems using a range of parent selection methods in a full-factorial design. Specifically, we applied each of tournament, random, fitness sharing, lexicase, and Eco-EA selection to the following problems: an exploration diagnostic, Count Odds, Logic-9, sorting networks, and NK Landscapes.

4.2.1 Selection Methods

4.2.1.1 Tournament

In tournament selection, T individuals are randomly selected from the population to form a "tournament". The individual in the tournament that has the highest fitness is selected to reproduce. Tournament selection is included here as a control, as it is simple and introduces no pressure for diversity.

In the exploration diagnostic experiments, we used a tournament size of $T = 8$ for consistency with [15]. In the other experiments, we used a tournament size of $T = 2$.

4.2.1.2 Random

Random selection is also included in this paper as a control. In this selection method, individuals are selected at random to reproduce.

4.2.1.3 Fitness Sharing

Fitness sharing is one of the earliest diversity maintenance techniques [10]. It is very similar to tournament selection, except that before the tournament occurs, the fitness of each individual is discounted in proportion to the number of similar individuals and the degree of their similarity. As a result, fitness sharing creates pressure for increased population diversity.

Fitness sharing requires two parameters: α (which controls the shape of the sharing function) and σ_{sh} (which controls the niche width). Here, we used $\alpha = 1$ and $\sigma_{sh} = 2$.

4.2.1.4 Lexicase Selection

Lexicase selection is designed to work in situations where fitness can be broken down into multiple constituent parts [28]. Conventionally, these constituent parts, or fitness criteria, are individual test cases in a genetic programming problem, but lexicase selection has been shown to work well in other scenarios too [3, 20]. To select a parent with lexicase selection, the fitness criteria are randomly ordered. Then the criteria are stepped through in order, with all but the best performers on each criterion being eliminated from consideration for this selection event. This process continues either until only one individual remains to be selected or until there are no criteria left, in which case an individual is randomly selected from those remaining. Lexicase selection is known to generally maintain high levels of diversity [12, 14], and preliminary evidence suggests that lexicase selection is also particularly effective at maintaining phylogenetic diversity [7].

4.2.1.5 Eco-EA

Like lexicase selection, Eco-EA is designed to work on problems that have various sub-components [8, 9]. The original inspiration for Eco-EA came out of problems in which solutions to multiple simpler tasks served as helpful building blocks for solving a more complex overall task. These building block tasks are analogous to fitness criteria in lexicase selection [3, 7]. Eco-EA associates each sub-task with a limited resource. Individuals that perform a task gain fitness while resources associated with that task are available; however, this decreases that task's associated resources. When the resource is depleted, there is no fitness benefit to performing the task. New resources flow in and out at a constant rate over the course of each update. As such, there is pressure for diversity.

The specific fitness gain associated with performing a task is described by the following function:

$$\text{total\_fitness} = \text{base\_fitness} * 2^{\min(\text{score}^{\alpha} * C_f * \text{resource, max\_bonus})} \tag{4.1}$$

where base fitness is the individual's fitness according to the global fitness function, score is the individual's score on the current sub-task, α is a variable that tunes the shape of the relationship between sub-task performance and fitness gain, C_f is the fraction of available resources that can be consumed, resource is the amount of the resource available, and max bonus is the maximum allowed bonus. To strengthen the negative frequency dependence, a cost can be applied for attempting to use more resource than is available.

For the exploration diagnostic, we used $C_f = 1$, $\alpha = .25$, cost =, 1, and max bonus = 5. Resources flow in at a rate of 250 units per generation. Every generation, 0.01% of the total quantity of each resource flows out. For the other experiments, we used $C_f = 0.01$, $\alpha = 2$, cost = 3, and max bonus = 5. Resources flow in at a rate of 50 units per generation and 5% of each resource flows out each generation.

4.2.2 *Problems*

4.2.2.1 Exploration Diagnostic

Recently, Hernandez et al. proposed a suite of "diagnostic" fitness landscapes designed to test evolutionary algorithms in a controlled environment [15]. Among these is an exploration diagnostic, which tests an algorithm's ability to simultaneously explore many different paths through a fitness landscape. In this diagnostic fitness landscape, genomes are vectors of floating point numbers. An individual's fitness is determined by finding the highest value in its genome, which represents the start of the values that contribute to fitness. This value and all subsequent monotonically decreasing values are summed together to produce the total fitness. For lexicase selection and Eco-EA, the fitness contribution of each individual site was used as the selection criterion or sub-task respectively.

Our hypotheses are predicated on the idea that phylogenetic diversity is a better measure of fitness landscape exploration than more conventionally used diversity metrics. As such, the exploration diagnostic represents an ideal context in which to test our hypotheses. Moreover, this fitness landscape provides a smooth and controlled environment with more intuitive diversity dynamics. In contrast, more complex fitness landscapes can yield idiosyncratic crashes in diversity that vary based on the path(s) the population takes. Thus, we will use the exploration diagnostic for our in-depth analysis, and provide results in the more complex landscapes as a follow-up investigation of the robustness of our findings.

4.2.2.2 Count Odds

Count Odds is part of a benchmark suite of program-synthesis problems derived from a set of real programming problems given to introductory computer science students [13]. These problems are specified using sets of testcases where each testcase contains inputs and expected outputs. Solutions to these problems consist of evolved code that produces the correct output for a given set of inputs. In the Count Odds problem, programs are given a list of numbers as input and must output an integer indicating how many of those numbers are odd.

For lexicase selection and Eco-EA, each test case is used as one selection criterion or sub-task, respectively. For fitness sharing, similarity is calculated based on the euclidean distance of the vector of performances on all test cases.

4.2.2.3 Logic-9

In Logic-9, a solution comprises code that, when executed, performs the nine one- and two-input bitwise Boolean logic tasks (*not*, *and*, *or*, *nand*, *orn*, *andn*, *nor*, *xor*, and *equ*) [5, 24]. To evaluate a program, we provide two numeric inputs that the program may use during its execution. Over the course of evaluation, a program may output numbers. We check whether each output is the result of performing one of the nine logic operations. If so, the program is counted as having completed that operation. Every logic operation completed increases the program's score by one. In addition to the Boolean logic tasks, programs get credit for solving the *echo* task, in which the program must output one of the numbers it received as input. This task is included because it is known to be helpful in scaffolding the evolution of the more complex logic tasks.

For lexicase selection and Eco-EA, each logic task is used as one selection criterion or sub-task, respectively. For fitness sharing, similarity is calculated based on the euclidean distance of the vector of which tasks were completed.

4.2.2.4 NK Landscapes

In NK Landscapes, individuals are bitstrings of length N [19]. Each position in the bitstring has an associated table used to look up the fitness contribution of that position, which is based on the value (0 or 1) at that position and the values of K neighboring positions. Thus, each lookup table comprises 2^{K+1} entries, one for each possible sequence of relevant bits. Total solution quality for a given bitstring individual is calculated by summing up the fitness contributions from each site.

An NK Landscape is generated by randomly selecting a value between 0 and 1 to be the fitness contribution for each entry in each site's fitness contribution lookup table. N determines the size of the fitness landscape, and K determines the landscape's level of epistasis or ruggedness (i.e., how interdependent the fitness contribution of each site is on other sites). For example, if K is 0, each site has two possible fitness contributions, one for if that position is set to 1 and one for if it is set to 0. If K is set to 1, each site has four possible fitness contributions based on the value of that site and the value of its neighbor (one each for 00, 01, 10, and 11). Every increase in K doubles the possible different fitness contributions.

For lexicase selection and Eco-EA, the fitness contribution from each site is used as one selection criterion or sub-task, respectively. For fitness sharing, similarity is based on the Hamming distance between the solutions. For the experiments presented here, we used $K = 3$ and $N = 20$.

4.2.2.5 Sorting Networks

Sorting networks are computational units designed to sort fixed-length sequences of numbers via a set of comparisons between pre-specified positions in the sequence [27]. When the network compares two positions, the numbers in them are swapped if they are out of order. In this problem, individuals are represented as sequences of comparators between positions. As with the Count Odds problem, fitness is assessed via a sequence of test cases. Once all test cases are solved correctly, individuals can receive additional fitness bonuses for having as few comparators as possible. For lexicase selection and Eco-EA, each test case is used as a single selection criterion or sub-task respectively.

Here, we evolved sorting networks to sort 30 values and test them on 100 test cases. The maximum allowed number of comparators per network was 128.

4.2.3 Computational Substrates

For the genetic programming problems (Count Odds and Logic-9), we evolved linear genetic programs where each genome is a sequence of simple computational instructions. Most notably, the instruction set is designed to support the evolution of modularity by supporting the encapsulation of subroutines into "scopes". Programs have a set of read-only memory spaces used to provide input and a set of write-only memory spaces to be used as output. Each instruction in the genome is executed in sequence. If execution reaches the end of the genome before the program runs out of evaluation time, execution will loop back around to the front of the genome. We propagated programs asexually and applied the following mutations to offspring, each with a probability of 0.005: (1) instruction substitutions, (2) point insertions and deletions, and (3) substituting the argument being supplied to an instruction. This linear genetic programming representation is described in more detail in [5].

For NK Landscapes, we evolved bitstrings. We propagated individuals asexually, and we applied bit flip mutations at a per-bit rate of 0.01.

In the exploration diagnostic, we evolved vectors of 50 floating point values where each value ranged between 0.0 and 25.0. We reduced the size of these ranges from those used by [15] to ensure that Eco-EA had sufficient time to solve the problem within our allocated computational budget. We used asexual reproduction, and each position had a 0.007 probability of mutating. Mutations modify a value by a number drawn from a normal distribution with mean 0 and standard deviation 1.

Sorting networks are represented as sequences of pairwise comparators. They can mutate via insertions of new comparators (0.0005 probability), duplications of existing comparators (0.0005 probability), deletions of existing comparators (0.001 probability), swapping pairs of existing comparators (0.001 probability), and substituting different indices in existing comparators (0.001 probability).

4.2.4 *Other Parameters*

For the exploration diagnostic, we used a population size of 500 and allowed runs to evolve for 500,000 generations. This length was selected to ensure that all selection schemes (most notably Eco-EA) had adequate time to find a good solution. For the other fitness landscapes, we used a population size of 1000 and allowed runs to evolve to 1,000 generations.

4.2.5 *Phylogenetic Diversity Metrics*

A wide variety of phylogenetic diversity metrics have been developed, and the extent to which they capture different information from each other is an area of active research [29]. All of them require that you keep track of the full phylogeny (ancestry tree) of a population. For further discussion of building phylogenies in the context of evolutionary computation, see [4]. Here, we focus on two classes of metrics: pairwise distance metrics and evolutionary distinctiveness metrics.

Pairwise distance metrics calculate the number of edges[1] in the shortest path between each pair of nodes associated with extant taxa (i.e., taxonomic units corresponding to individuals in the current population) [31]. The resulting set of distances can then be summarized by calculating its minimum, maximum, mean, or variance. Each of these statistics produces a different phylogenetic diversity metric with different properties.

Evolutionary distinctiveness metrics assign an evolutionary distinctiveness score to each extant taxon [16]. This score takes into account each branch's age, and represents how evolutionarily distant each taxon is from all other extant taxa. To calculate evolutionary distinctiveness, the age of each branch is calculated and divided by the number of extant taxa the branch ultimately leads to. A taxon's evolutionary distinctiveness is the sum of the values calculated for all branches between that taxon and

[1] Weighted edges can also be used, in which case the weights along the path should be summed. Here, we use unweighted edges.

the tree's root. As with pairwise distances, the set of evolutionary distinctiveness scores can be summarized by taking its minimum, maximum, mean, or variance.

4.2.6 Analysis Techniques

4.2.6.1 Statistics

Correlations among variables at a fixed time point were measured using Spearman correlations. We used Spearman correlations rather than Pearson correlations due to the fact that many of the relationships being measured were non-linear (but still monotonically increasing). To compare different conditions, we used Kruskal-Wallis tests with subsequent pairwise Wilcoxon rank-sum tests and a Bonferonni correction for multiple comparisons. To ensure statistical rigor, we decided what set of statistical comparisons to perform based on analysis of an initial pilot data-set. We then re-ran all experiments with different random seeds and performed the pre-determined analysis on this new data to generate the results presented here.

4.2.6.2 Transfer Entropy

As alluded to previously, there is a positive feedback loop between diversity and evolutionary success during the initial phase of evolution. To further complicate matters, fully solving a problem can lead diversity to crash as that solution sweeps through the population. For these reasons, looking at correlations between fitness and diversity at any given time point gives us only limited information about their relationship. One approach to getting around this problem is to take all measurements at the time step when a perfect solution is first discovered [6]. While this approach helps, it does not address the question of what is driving the feedback loop. Moreover, even before a perfect solution evolves, the evolution of partial solutions may initiate partial selective sweeps.

In order to understand the specific role that different types of diversity play in driving the feedback loop between diversity and success, we turn to an analytical approach called causality analysis. As the nature of causality can quickly drift into murky philosophical territory, here we specifically use a notion of causality called Granger causality [1, 11]. We say that X Granger-causes Y if past values of X contain information about the current value of Y above and beyond the information that past values of Y contain about the current value of Y. This definition comes from the insights that 1) the past causes the future, not the other way around, and 2) if X and Y are jointly caused by an external process, that process will be captured by the information that past values of Y have about the current value of Y.

Granger causality is normally measured in the context of vector auto-regressive of models. However, our data do not match the assumptions of such models (particularly stationarity). Thus, we measure Granger-causality with an information theoretic metric called Transfer Entropy [26, 33]. In information theoretic terms, transfer entropy is $I(Y_t; X_{t-k}|Y_{t-k})$, the conditional mutual information between Y_t and $X_{t-k}|Y_{t-k}$. Here, Y is the variable being predicted, t is the time point it is being predicted at, X is the variable we are using the predict Y, and k is the "lag". The lag specifies the time scale on which we are interested in measuring Granger causality. In other words, it indicates which past value of X we are attempting to use to predict the current value of Y.

Often, the goal of these measurements is simply to establish the direction of Granger-causality. In those cases, very short lags are often used. In this case, however, it would also be valuable to know whether predictive capability is maintained over large lags. If it were, there would be a variety of useful practical implications. For example, we could potentially use the diversity at a relatively early time point to predict whether a given run of evolutionary computation will be successful. For this reason, we measure Transfer Entropy using lags ranging from 10 generations to 100,000 generations.

Here, we measure both the Transfer Entropy between fitness and phylogenetic diversity and the Transfer Entropy between fitness and phenotypic diversity. Note that, because Transfer Entropy is a value calculated based on two time series, we cannot meaningfully lump multiple replicates into the same calculation. Thus, for each condition we will end up with a distribution of Transfer Entropy values.

4.2.7 Code Availability

All code used in this paper is open source and freely available in the supplemental material [2]. Research code was written in C++ using the Empirical library [23]. Data analysis was performed using the R statistical computing language, version 4.0.4 [25], the ggplot2 [32], ggpubr [18], and infotheo [21] libraries.

A C++ implementation of phylogeny tracking and all phylodiversity metrics used here is available in the Empirical library [23]. This implementation is designed to plug into any computational evolution code written in C++.

4.3 Results and Discussion

4.3.1 Do Phylogenetic Metrics Provide Novel Information?

If phylogenetic diversity is to tell us anything useful, a necessary first step is that it provide information that more commonly-used metrics do not. It is particularly important to establish this distinction in light of the fact that phylogenetic metrics are typically more computationally expensive to calculate.

First, we investigated the relationships between different metrics of phenotypic and phylogenetic diversity. The two measurements of phenotypic diversity that we analyzed were phenotypic richness and phenotypic Shannon diversity, the two most commonly used diversity metrics (see supplemental material [2]). Unsurprisingly, phenotypic richness and phenotypic Shannon diversity are closely correlated across all conditions. We performed the rest of the analyses in this paper using both richness and Shannon diversity, but observed qualitatively the same results for both. For simplicity, we only present the results using richness (for Shannon diversity results, see supplemental material [2]).

There are many ways of measuring phylogenetic diversity [29]. Here, we focus on the pairwise distance and evolutionary distinctiveness metrics (see Sect. 4.2.5 for more information). Minimum pairwise distance was not informative, as there are nearly always at least two taxa a distance of 1 away from each other. In the conditions observed here, the other pairwise distance metrics tended to correlate fairly closely with each other. Correlations among evolutionary distinctiveness metrics were weaker and less consistent, but still present in most cases. In contrast, we did not observe consistent relationships between pairwise distance metrics and evolutionary distinctiveness metrics. While there were sometimes strong correlations within a condition, the direction of these correlations varied. For this reason, we performed subsequent analyses using both mean pairwise distance and mean evolutionary distinctiveness. As we observed qualitatively the same results for both, here we present only the results using mean pairwise distance (for evolutionary distinctiveness results, see supplemental material [2]).

Next, we compared the phenotypic metrics to the phylodiversity metrics. Intuitively, we might expect them to be highly correlated. In practice, however, we see that the correlation between phenotypic and phylogenetic diversity is not consistently[2] significantly different from 0 (see Fig. 4.2). In some cases, the correlation is even negative.

As previously noted, diversity metrics can be sensitive to the exact time at which they are measured. To confirm that this lack of instantaneous correlation is indicative of a consistent lack of relationship, we plotted phenotypic and phylogenetic diversity over time (see Figs. 4.3 and 4.4). Indeed, phenotypic and phylogenetic diversity behave differently over long temporal scales as well.

Having established that phenotypic and phylogenetic diversity are not reliably correlated, we next asked whether either diversity metric can predict the other. Based on our hypothesis that phylogenetic diversity is useful because it more directly indicates a population's spread across the fitness landscape, we might expect current phylogenetic diversity to predict future phenotypic diversity. We tested this hypothesis by measuring the transfer entropy from phylogenetic diversity to phenotypic diversity. Consistent with our hypothesis, transfer entropy from phylogenetic diversity to phenotypic diversity was generally higher than transfer entropy from phenotypic

[2] The correlation for tournament selection in the exploration diagnostic is incredibly high, however (1) the observed range of mean pairwise distance is so low that the correlation is almost certainly an artifact, and (2) this correlation is not observed for other fitness landscapes.

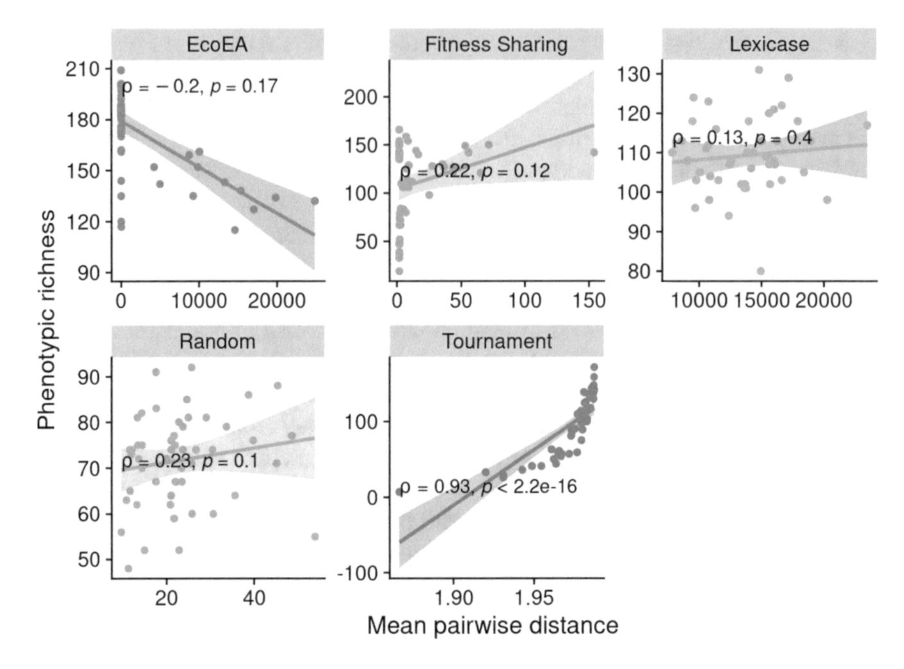

Fig. 4.2 Relationship between phenotypic diversity and phylogenetic diversity at a single time point for the exploration diagnostic. Values were measured at the final time point (generation 500,000). Spearman correlation coefficients are shown for each selection scheme, along with a 95% confidence interval around regression lines. Note that axis scales vary between panels

diversity to phylogenetic diversity (see Fig. 4.5). Thus, in the Granger sense of causality, we can say that phylogenetic diversity produces phenotypic diversity to a greater extent than phenotypic diversity produces phylogenetic diversity.

In the exploration diagnostic fitness landscape, the only exceptions to this observation are our two control selection schemes: tournament selection and random selection. In tournament selection, neither form of diversity is particularly predictive of the other. This behavior is unsurprising, as tournament selection generally maintains minimal levels of both forms of diversity. In random selection, both forms of diversity are somewhat predictive of each other. In the other fitness landscapes we observe more variation in transfer entropy (see supplemental material [2]).

Taken together, these results provide strong evidence that, in an evolutionary computation context, phylogenetic diversity metrics provide information that phenotypic diversity metrics do not. We base this conclusion on the lack of consistent correlation between these metrics at a fixed point in time, the differences in their long-term behavior, and that fact that the transfer entropy from phylogenetic diversity to phenotypic diversity is higher than the other way around (implying that phylogenetic diversity contains information about future phenotypic diversity that current phenotypic diversity does not contain about future phylogenetic diversity). We may now proceed to

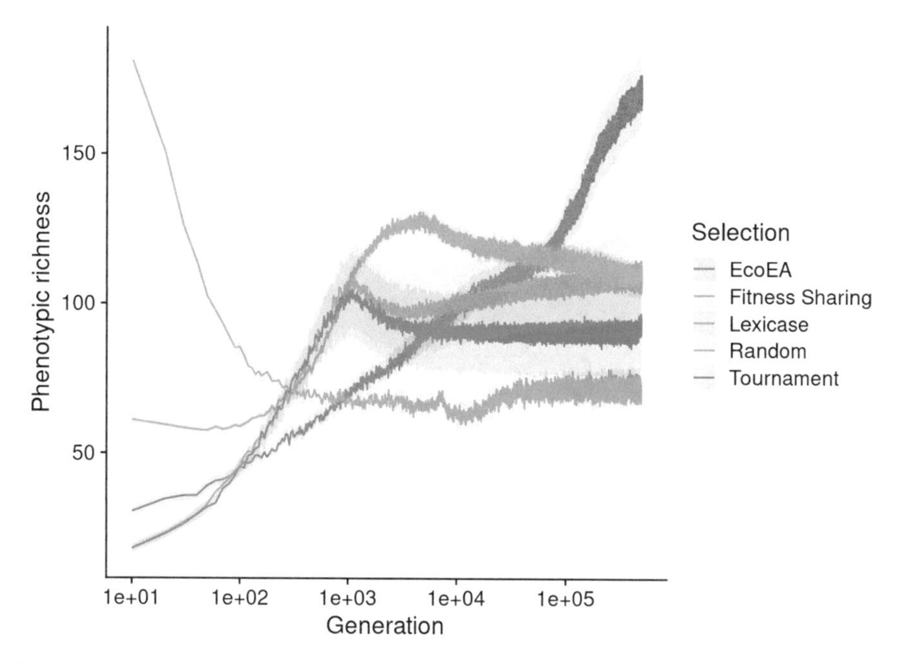

Fig. 4.3 Phenotypic richness over time for each selection scheme on the exploration diagnostic. Shaded areas represent 95% confidence interval around the mean for all 50 replicates for each selection scheme. Note that the x-axis is on a log scale

ask whether that information is actually useful for understanding problem-solving success in evolutionary computation.

4.3.2 Do Phylogenetic Metrics Predict Problem-Solving Success?

Having established that phylogenetic metrics provide different information than phenotypic diversity metrics, the next question to ask is what that information can tell us. In the exploration diagnostic landscape, we can see some intuitive connections between both types of diversity and fitness (see Fig. 4.6). Excluding random selection, the final performance of a selection scheme appears to be correlated with the final level of phylogenetic diversity maintained by that selection scheme (but not with the level of phenotypic diversity). In the other fitness landscapes, however, the connection between a selection scheme's ability to maintain diversity of either type and its ability to succeed is less obvious (see supplemental material [2]). This increased complexity is unsurprising, as success in the exploration diagnostic landscape is based primarily on the ability to explore; succeeding on the other fitness landscapes is more complicated.

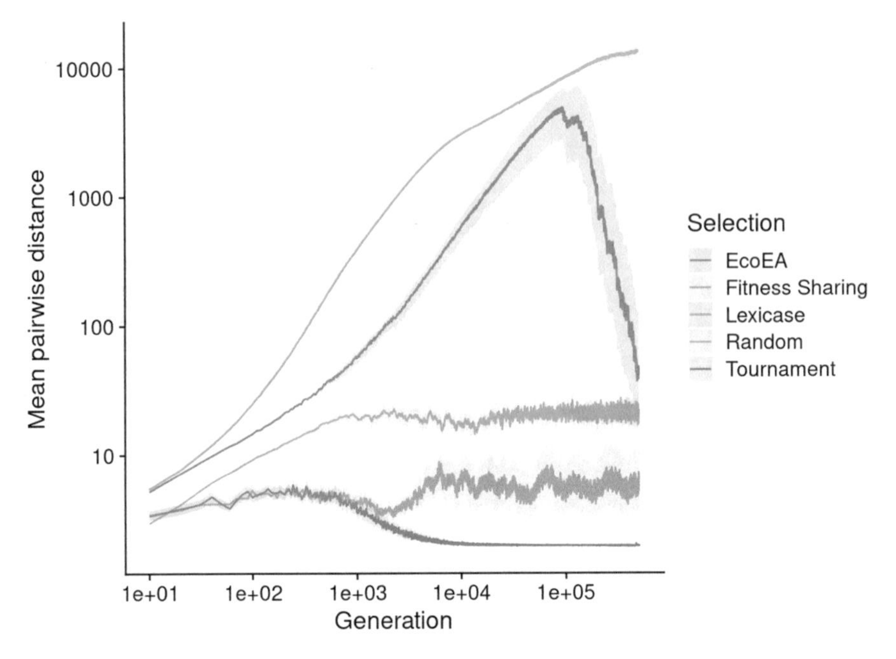

Fig. 4.4 Phylogenetic diversity (mean pairwise distance) over time for each selection scheme on the exploration diagnostic. Shaded areas represent 95% confidence interval around the mean for all 50 replicates for each selection scheme. Note that the x-axis is on a log scale

It should also be noted that, while there is a correlation between phylogenetic diversity and fitness in the aggregate on the exploration diagnostic landscape, there is not a consistent correlation among individual runs within a selection scheme (see supplemental material [2]). Eco-EA, for instance, appears to lose phylogenetic diversity as it approaches a good solution, leading to a negative correlation between fitness and phylogenetic diversity. For most of the other selection schemes, there is no significant correlation.[3]

To understand the precise dynamics driving the relationship between diversity and success, we measured the transfer entropy of phylogenetic diversity to fitness and the transfer entropy of phenotypic diversity to fitness. In the exploration diagnostic landscape, we see that, for the three non-control selection schemes, phylogenetic diversity is substantially more predictive of future fitness than phenotypic diversity (see Fig. 4.7). The predictive power of both types of diversity weakens substantially in the other fitness landscapes (see Fig. 4.8). However, when there is a discernible difference, phylogenetic diversity has higher transfer entropy than phenotypic diversity.

[3] In the pilot data set, we observed a strong positive correlation between phylogenetic diversity and fitness for lexicase selection. However, this correlation disappeared when we re-ran the experiments to generate the final data set.

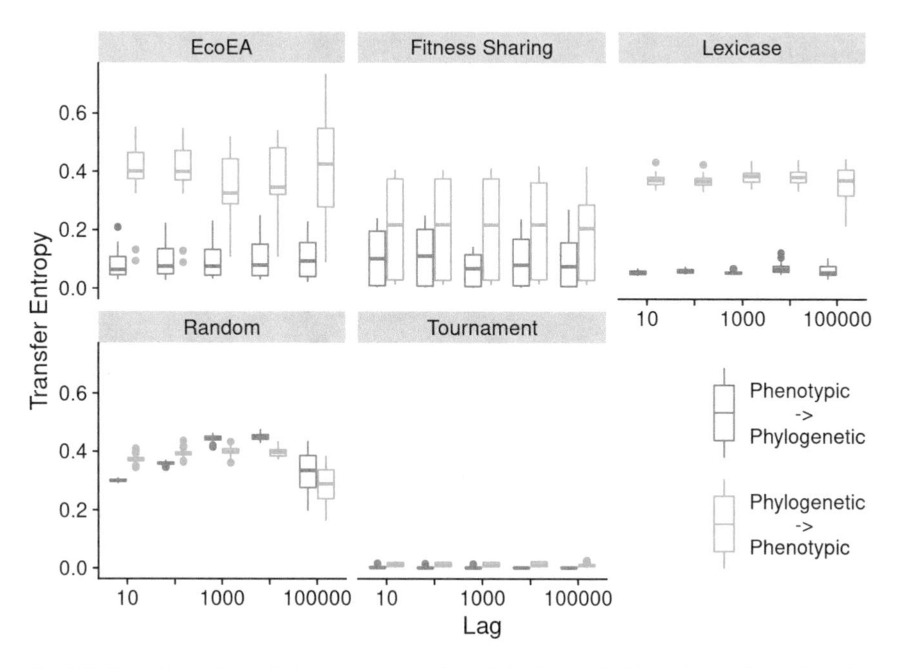

Fig. 4.5 Transfer entropy between phenotypic and phylogenetic diversity on the exploration diagnostic. Each boxplot shows the distribution of observed transfer entropies for each direction of transfer, lag, and selection scheme. Results shown here are for 50 replicate runs of each selection scheme on the exploration diagnostic. Note that the x-axis is on a log scale

From these results, we can conclude that, for selection schemes that maintain diversity, phylogenetic diversity "Granger-causes" success to a greater extent than phenotypic diversity does. We interpret this as strong evidence that phylogenetic diversity is, in general, more predictive of success than phenotypic diversity (as measured by phenotypic richness and phenotypic Shannon diversity).

4.4 Conclusion

We have demonstrated that, in the context of evolutionary computation, phylogenetic diversity metrics capture information information that is substantially different from the information captured by conventionally used phenotypic diversity metrics (phenotypic richness and phenotypic Shannon diversity). The extent of this difference appears to vary by problem and by selection scheme, but it is evident in (1) the lack of consistent strong correlation between phenotypic and phylogenetic diversity at a fixed point in time, (2) the lack of similarity in long-term trends in phenotypic and phylogenetic diversity, and (3) the fact that phylogenetic diversity is better able to predict future phenotypic diversity than the other way around.

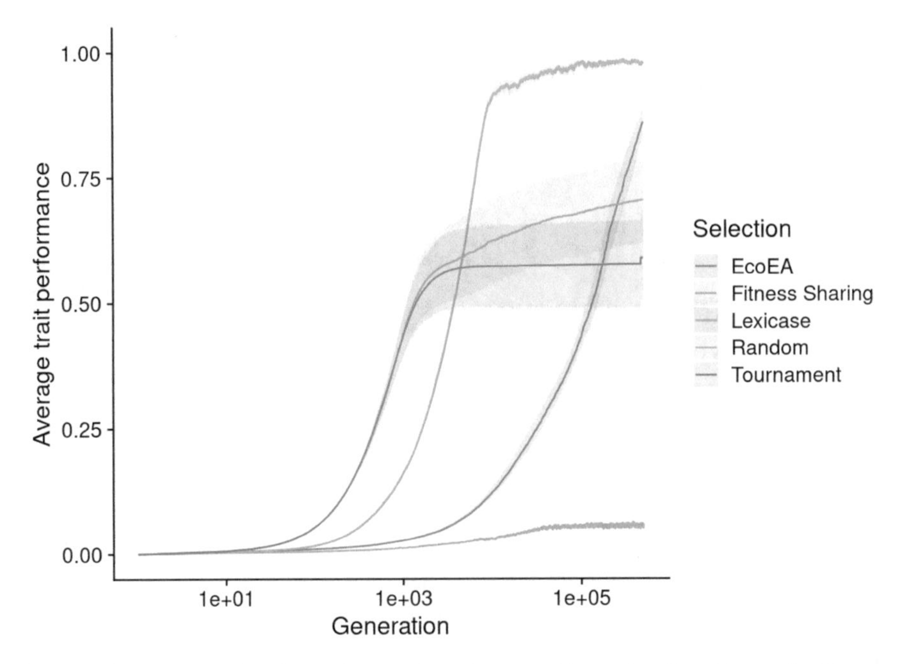

Fig. 4.6 Fitness over time for each selection scheme on the exploration diagnostic. Shaded areas represent 95% confidence interval around the mean for all 50 replicates for each selection scheme. Note that the x-axis is on a log scale. The y-axis shows the proportion of the maximum possible fitness achieved

Moreover, our results also suggest that phylogenetic diversity is generally a stronger driver of success than phenotypic diversity (in an evolutionary computing context). This finding also varies by selection scheme and by problem, with selection schemes that maintain diversity showing a stronger relative effect of phylogenetic diversity than tournament selection does. Impressively, this is true even at very long time lags; phylogenetic diversity provides predictive information about fitness tens of thousands of generations in the future.

Taken together, these results suggest that it may be worthwhile for researchers studying diversity in evolutionary computation to measure phylogenetic diversity in addition to or instead of phenotypic diversity. Doing so will take us a step closer to identifying diversity that is helpful to solving a given problem. Additionally, these results hint at the possibility of using phylogenetic diversity early in a run of evolutionary computation as a predictor of which runs will go on to be most successful. Anecdotally, we have found phylogenetic diversity in just the first few generations to be a useful indicator of whether we have correctly selected parameters for fitness sharing and Eco-EA.

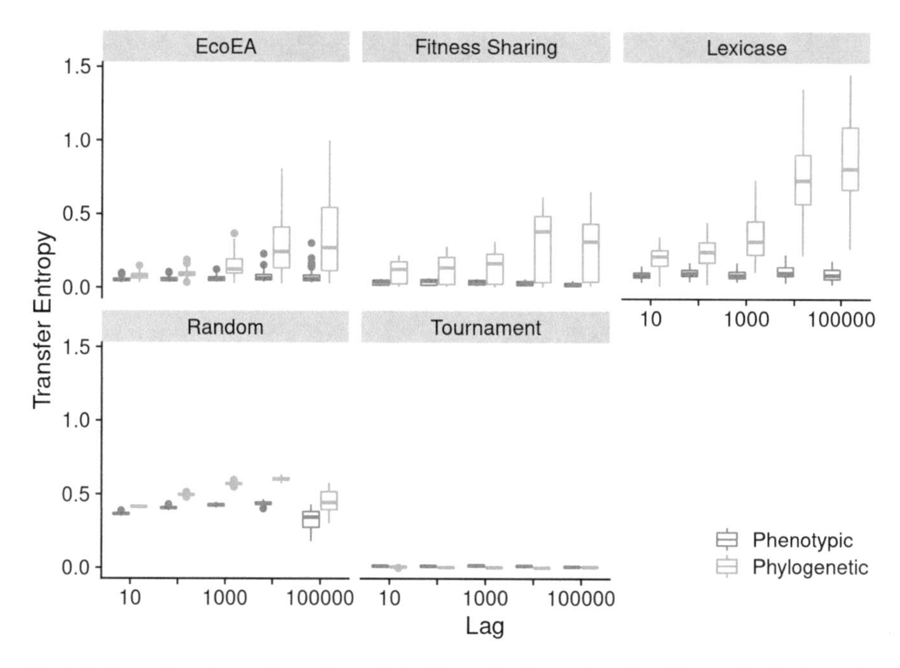

Fig. 4.7 Transfer entropy from diversity to fitness for the exploration diagnostic. Each boxplot shows the distribution of observed transfer entropies for each type of diversity, lag, and selection scheme. Results shown here are for 50 replicate runs of each selection scheme on the exploration diagnostic. Note that the x-axis is on a log scale

In this paper, we have barely scratched the surface of phylogenetic metrics that may be relevant to evolutionary computation. Given our findings thus far, a more thorough investigation of other phylogeny metrics in the context of evolutionary computation is warranted. Additionally, future work should evaluate the relationships between other conventionally used approaches to measuring diversity (e.g., genetic diversity). As we improve our understanding of these relationships, we may even be able to use them to make inferences about fitness landscapes [4]. We hope that the results presented here will inspire others to incorporate a phylogenetic perspective into their evolutionary computation research.

4.5 Author Contributions

ED conceptualized the questions and experiments in this chapter, wrote the code for the non-exploration-diagnostic experiments, ran all experiments, analyzed the data, and wrote the first draft of this chapter. JGH and AL wrote the code for the exploration diagnostic experiments and assisted with the data analysis and writing.

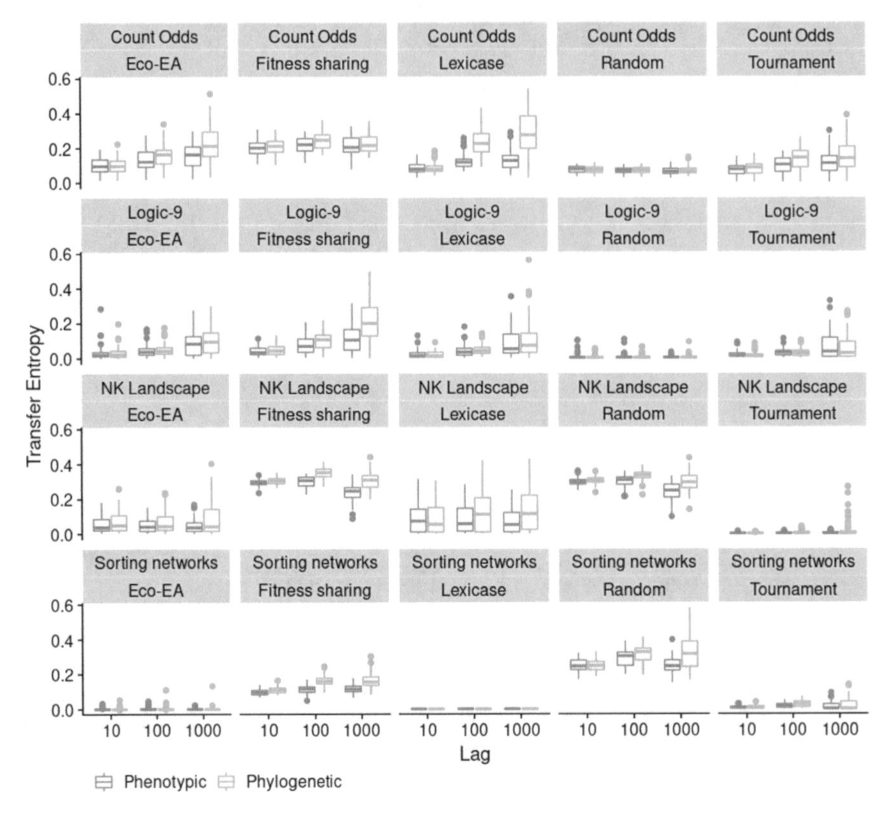

Fig. 4.8 Transfer entropy from diversity to fitness for complex fitness landscapes. Each boxplot shows the distribution of observed transfer entropies for each type of diversity, lag, and selection scheme. Results shown here are for 50 replicate runs of each selection scheme on each landscape. Note that the x-axis is on a log scale

Acknowledgements We thank members of the MSU ECODE lab, the MSU Digital Evolution lab, and the Cleveland Clinic Theory Division for the conversations that inspired this work. This research was supported by the National Science Foundation (NSF) through the BEACON Center (Cooperative Agreement DBI-0939454). Michigan State University provided computational resources through the Institute for Cyber-Enabled Research. Any opinions, findings, and conclusions or recommendations expressed in this material are those of the author(s) and do not necessarily reflect the views of the NSF, UM, or MSU.

References

1. Bressler, S.L., Seth, A.K.: Wiener—granger causality: a well established methodology. NeuroImage **58**(2), 323–329 (2011)
2. Dolson, E.: Supplemental material for "What can phylogenetic metrics tell us about useful diversity in evolutionary algorithms?" at GPTP 2021 (2021). https://doi.org/10.5281/zenodo.4733407

3. Dolson, E., Banzhaf, W., Ofria, C.: Applying ecological principles to genetic programming. In: Banzhaf, W., Olson, R.S., Tozier, W., Riolo, R. (eds.) Genetic Programming Theory and Practice XV, pp. 73–88. Springer International Publishing, Cham (2018)
4. Dolson, E., Lalejini, A., Jorgensen, S., Ofria, C.: Interpreting the tape of life: ancestry-based analyses provide insights and intuition about evolutionary dynamics. Artif. Life **26**(1), 1–22 (2020)
5. Dolson, E., Lalejini, A., Ofria, C.: Exploring genetic programming systems with map-elites. In: Banzhaf, W., Spector, L., Sheneman, L. (eds.) Genetic Programming Theory and Practice XVI, pp. 1–16. Springer International Publishing, Cham (2019)
6. Dolson, E., Perez, S., Olson, R., Ofria, C.: Spatial resource heterogeneity increases diversity and evolutionary potential. bioRxiv (2017). https://doi.org/10.1101/148973
7. Dolson, E.L., Banzhaf, W., Ofria, C.: Ecological theory provides insights about evolutionary computation. Peer J Preprints **6**, e27,315v1 (2018)
8. Goings, S., Goldsby, H.J., Cheng, B.H., Ofria, C.: An ecology-based evolutionary algorithm to evolve solutions to complex problems. Artif. Life **13**, 171–177 (2012)
9. Goings, S., Ofria, C.: Ecological approaches to diversity maintenance in evolutionary algorithms. In: IEEE Symposium on Artificial Life, 2009. ALife '09, pp. 124–130 (2009)
10. Goldberg, D.E., Richardson, J., Grefenstette, J.J.: Genetic algorithms with sharing for multimodal function optimization. In: Genetic algorithms and their applications: Proceedings of the Second International Conference on Genetic Algorithms, pp. 41–49. Lawrence Erlbaum, Hillsdale, NJ (1987)
11. Granger, C.W.J.: Investigating causal relations by econometric models and cross-spectral methods. Econometrica **37**(3), 424–438 (1969)
12. Helmuth, T., McPhee, N.F., Spector, L.: Lexicase selection for program synthesis: a diversity analysis. In: Riolo, R., Worzel, W.P., Kotanchek, M., Kordon, A. (eds.) Genetic Programming Theory and Practice XIII, Genetic and Evolutionary Computation, pp. 151–167. Springer International Publishing (2016)
13. Helmuth, T., Spector, L.: General program synthesis benchmark suite. In: Proceedings of the 2015 Annual Conference on Genetic and Evolutionary Computation, GECCO '15, pp. 1039–1046. ACM, New York, NY, USA (2015)
14. Helmuth, T., Spector, L., Matheson, J.: Solving uncompromising problems with lexicase selection. IEEE Trans. Evol. Comput. **19**(5), 630–643 (2015)
15. Hernandez, J.G., Lalejini, A., Ofria, C.: An Exploration of exploration: measuring the ability of lexicase selection to find obscure pathways to optimality (2021). arXiv:2107.09760 [cs]
16. Isaac, N.J.B., Turvey, S.T., Collen, B., Waterman, C., Baillie, J.E.M.: Mammals on the EDGE: conservation priorities based on threat and phylogeny. PLOS ONE **2**(3), e296 (2007)
17. Jackson, D.: Promoting Phenotypic Diversity in Genetic Programming. In: Schaefer, R., Cotta, C., KoÅodziej, J., Rudolph, G. (eds.) Parallel Problem Solving from Nature, PPSN XI. Lecture Notes in Computer Science, pp. 472–481. Springer, Berlin, Heidelberg (2010)
18. Kassambara, A.: ggpubr: 'ggplot2' Based Publication Ready Plots (2020). https://CRAN.R-project.org/package=ggpubr. R package version 0.4.0
19. Kauffman, S., Levin, S.: Towards a general theory of adaptive walks on rugged landscapes. J. Theor. Biol. **128**(1), 11–45 (1987)
20. Metevier, B., Saini, A.K., Spector, L.: Lexicase selection beyond genetic programming. In: Banzhaf, W., Spector, L., Sheneman, L. (eds.) Genetic Programming Theory and Practice XVI, Genetic and Evolutionary Computation, pp. 123–136. Springer International Publishing, Cham (2019)
21. Meyer, P.E.: Infotheo: information-theoretic measures (2014). https://CRAN.R-project.org/package=infotheo. R package version 1.2.0
22. Mouret, J., Doncieux, S.: Overcoming the bootstrap problem in evolutionary robotics using behavioral diversity. In: IEEE Congress on Evolutionary Computation, 2009. CEC'09, pp. 1161–1168. IEEE (2009)
23. Ofria, C., Dolson, E., Lalejini, A., Fenton, J., Jorgensen, S., Miller, R., Moreno, M.A., Stredwick, J., Zaman, L., Schossau, J., Gillespie, L., G, N.C., Vostinar, A.: Empirical (2018). https://doi.org/10.5281/zenodo.1439475

24. Ofria, C., Wilke, C.O.: Avida: a software platform for research in computational evolutionary biology. Artif. Life **10**(2), 191–229 (2004)
25. Team, R.C.: R: a language and environment for statistical computing. In: R Foundation for Statistical Computing, Vienna, Austria (2021). https://www.R-project.org/
26. Schreiber, T.: Measuring information transfer. Phys. Rev. Lett. **85**(2), 461–464 (2000)
27. Sekanina, L., Bidlo, M.: Evolutionary design of arbitrarily large sorting networks using development. Genet. Program. Evolvable Mach. **6**(3), 319–347 (2005)
28. Spector, L.: Assessment of problem modality by differential performance of lexicase selection in genetic programming: a preliminary report. In: Proceedings of the 14th Annual Conference Companion on Genetic and Evolutionary Computation, pp. 401–408. ACM (2012)
29. Tucker, C.M., Cadotte, M.W., Carvalho, S.B., Davies, T.J., Ferrier, S., Fritz, S.A., Grenyer, R., Helmus, M.R., Jin, L.S., Mooers, A.O., Pavoine, S., Purschke, O., Redding, D.W., Rosauer, D.F., Winter, M., Mazel, F.: A guide to phylogenetic metrics for conservation, community ecology and macroecology. Biol. Rev. **92**(2), 698–715 (2017)
30. Tucker, C.M., Davies, T.J., Cadotte, M.W., Pearse, W.D.: On the relationship between phylogenetic diversity and trait diversity. Ecology **99**(6), 1473–1479 (2018)
31. Webb, C.O., Ackerly, D.D., McPeek, M.A., Donoghue, M.J.: Phylogenies and community ecology. Annu. Rev. Ecol. Syst. **33**(1), 475–505 (2002)
32. Wickham, H.: ggplot2: Elegant Graphics for Data Analysis. Springer, New York (2016)
33. Yao, C.Z., Li, H.Y.: Effective transfer entropy approach to information flow among EPU, investor sentiment and stock market. Front. Phys. **8**, 206 (2020)

Chapter 5
An Exploration of Exploration: Measuring the Ability of Lexicase Selection to Find Obscure Pathways to Optimality

Jose Guadalupe Hernandez, Alexander Lalejini, and Charles Ofria

Abstract Parent selection algorithms (selection schemes) steer populations through a problem's search space, often trading off between exploitation and exploration. Understanding how selection schemes affect exploitation and exploration within a search space is crucial to tackling increasingly challenging problems. Here, we introduce an "exploration diagnostic" that diagnoses a selection scheme's capacity for search space exploration. We use our exploration diagnostic to investigate the exploratory capacity of lexicase selection and several of its variants: epsilon lexicase, down-sampled lexicase, cohort lexicase, and novelty-lexicase. We verify that lexicase selection out-explores tournament selection, and we show that lexicase selection's exploratory capacity can be sensitive to the ratio between population size and the number of test cases used for evaluating candidate solutions. Additionally, we find that relaxing lexicase's elitism with epsilon lexicase can further improve exploration. Both down-sampling and cohort lexicase—two techniques for applying random sub-sampling to test cases—degrade lexicase's exploratory capacity; however, we find that cohort partitioning better preserves lexicase's exploratory capacity than down-sampling. Finally, we find evidence that novelty-lexicase's addition of novelty test cases can degrade lexicase's capacity for exploration. Overall, our findings provide hypotheses for further exploration and actionable insights and recommendations for using lexicase selection. Additionally, this work demonstrates the value of selection scheme diagnostics as a complement to more conventional benchmarking approaches to selection scheme analysis.

J. G. Hernandez (✉) · A. Lalejini · C. Ofria
Michigan State University, East Lansing, MI, USA
e-mail: herna383@msu.edu

A. Lalejini
e-mail: lalejini@umich.edu

C. Ofria
e-mail: ofria@msu.edu

W. Banzhaf et al. (eds.), *Genetic Programming Theory and Practice XVIII*,
Genetic and Evolutionary Computation,
https://doi.org/10.1007/978-981-16-8113-4_5

5.1 Introduction

Lexicase-based parent selection algorithms have proven to be highly successful for finding effective solutions to test-based problems in genetic programming (GP) [10, 15, 34]. Lexicase selection's success is rooted in its ability to balance strong search space exploration with simultaneous exploitation. That is, lexicase selection maintains meaningfully diverse populations [12, 14] by promoting the coexistence of subpopulations that are each focused on different aspects of a problem (e.g., on different test cases or selection criteria) [5]. As such, lexicase selection algorithms are able to explore many promising problem-solving pathways in parallel, optimizing each until an overall solution is found.

Many genetic programming problems are multi-faceted where the quality of a candidate solution must be measured according to its performance on a set of test cases. For such problems, we must decide how to combine performances across many test cases in order to select promising individuals to produce offspring for the next generation. Traditional parent selection algorithms assess the quality of an individual by aggregating their performance on all test cases. The lexicase selection algorithm, however, chooses each parent based on the relative performances of candidate solutions on random permutations of the test set. Specifically, each time a parent is needed, the entire population is considered as candidates for selection, and the full set of test cases are shuffled; each test case is applied sequentially (in the given shuffled order) to the current set of candidates, removing all but the best candidates from consideration until only a single individual remains to be selected [18]. Because the ordering of test cases is different for each parent selection event, individuals that perform well on different subsets of problems are able to coexist [5]. Moreover, lexicase selection exerts strong selection pressure to optimize each subpopulation, as only the best candidates on different sequences of test cases are selected.

Indeed, the successes of the original lexicase selection algorithm have inspired numerous variants, each either specialized for solving different categories of problems or designed to address potential shortcomings of the original lexicase algorithm (e.g., computational efficiency). Such variants include epsilon lexicase [24, 25], down-sampled lexicase [19], novelty-lexicase [22], ALPS lexicase [10], and batch-lexicase selection [1]. Many of these variants have been rigorously benchmarked on their problem-solving success and on their ability to maintain phenotypic and phylogenetic diversity [7, 12, 13, 37]. However, benchmarking is often performed in the context of a particular GP system and with the overall goal of measuring performance on challenging computational problems (e.g., program synthesis benchmark problems from [11, 15]). While such benchmarking is critical for understanding the real-world applicability of a selection scheme, the specific problems used do not always allow us to disentangle the particular pros and cons of each scheme [21]. For this paper, we focus on one important aspect of lexicase-based selection schemes: How do we isolate the *exploration* capabilities of lexicase selection and its variants?

We introduce an "exploration diagnostic" and use it to test how well a set of parent selection algorithms can explore a simple landscape with many uphill pathways of differing peak fitnesses. Our exploration diagnostic allows for the total number of possible evolutionary pathways to be tuned, enabling practitioners to find where an algorithm's exploratory abilities begin to fall off. First, we verify established expectations that lexicase selection better facilitates search space exploration than tournament selection, a more traditional selection algorithm. Next, we evaluate lexicase selection on our exploratory diagnostic with an increasing number of possible pathways identify its exploratory limitations. Finally, we apply our exploration diagnostic to four variants of lexicase selection: epsilon lexicase, down-sampled lexicase, cohort lexicase, and novelty-lexicase selection.

We find that lexicase selection drives performance improvement at each of the exploration diagnostic difficulty levels that we evaluated. Lexicase selection finds nearly perfect solutions for fitness landscapes with a small number of pathways to be explored, and performance gradually declines as the number of possible evolutionary pathways increases. Additionally, we show that lexicase selection can be sensitive to the ratio between population size and the number of test cases used for evaluating candidate solutions. For small values of ε, epsilon lexicase improves the exploratory capacity of lexicase selection. Random subsampling via either down-sampled or cohort lexicase degrades exploratory capacity, but cohort partitioning better preserves lexicase's exploratory capacity than down-sampling. Finally, we did not find compelling evidence that novelty-lexicase improves performance on the exploration diagnostic relative to standard lexicase selection; in fact, the addition of novelty test cases can substantially degrade lexicase's diagnostic performance.

5.2 Exploration Diagnostic

Understanding how parent-selection algorithms affect exploration and exploitation within a search space is crucial to tackling increasingly challenging problems. This information can help determine what modifications to an evolutionary algorithm may be needed to improve the likelihood of finding a high quality solution. Different selection schemes (or other components of an evolutionary algorithm) can alter the trade-off between exploitation and exploration [6]. An exploitation-only selection scheme will push the population to the closest optimum and not allow it to explore other promising regions of the search space. Conversely, an exploration-only selection scheme will scatter the population across the entire search space but is unlikely to reach nearby optima. Hence, striking a balance between exploration and exploitation is critical to finding high-quality solutions. Here, we introduce a diagnostic that challenges selection schemes to explore multiple avenues of a search space, each with an upward pathway, with the goal of finding the best avenue to hill climb.

We balanced both exploitation and exploration in our diagnostic. Specifically, we designed a problem with many upward pathways that all have identical slopes, but vary in total length. Since shorter pathways are always equivalent to the beginning of

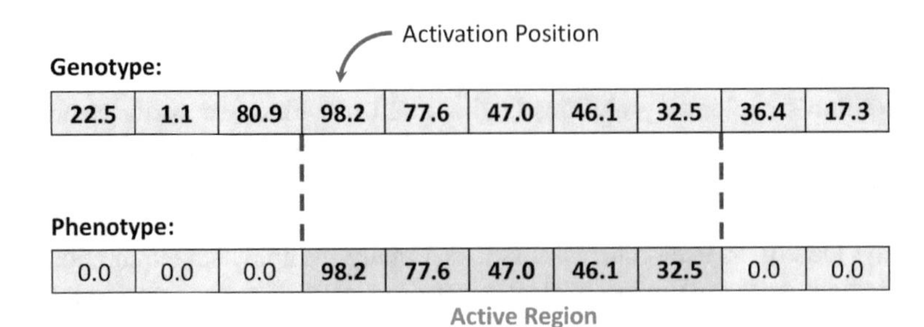

Fig. 5.1 An example evaluation with the exploration diagnostic. A candidate solution with a cardinality of 10 is analyzed. The highest value in its vector is identified as 98.2, and its position is marked as the beginning of the active region. The next four values are all in a decreasing sequence (77.6, 47.0, 46.1, and 32.5) and are thus all considered part of the active region. The value after that (36.4) is greater than its predecessor and thus left inactive, closing the active region. All values not in the active region are expressed in the phenotype as 0.0. The total fitness of the sequence is the sum of the values in the phenotype or $0.0 + 0.0 + 0.0 + 98.2 + 77.6 + 47.0 + 46.1 + 32.5 + 0.0 + 0.0 = 301.4$

longer pathways, exploration is critical for finding the longest pathway (which will lead to the global optimum). In the end, the only way for an evolving population to determine the length of a pathway is to follow it.

Candidate solutions for this diagnostic are numerical vectors of a designated size (its "cardinality"—we used 100 as the default cardinality in this work). Cardinality determines the number of pathways to local optima in the fitness landscape. Each value in a candidate solution is a floating-point number between 0.0 and 100.0. To evaluate a candidate solution, we first scan its vector to find the maximum value and designate its position as the "activation position" for calculating its fitness. From an intuitive perspective, the activation position defines which peak the candidate solution is climbing toward. Beginning at the activation position, we sum all consecutive values that are less than or equal to each previous position. We stop when either a position is no longer monotonically non-increasing or we reach the end of the vector. We refer to this consecutive sequence of scored values as the "active region" of the candidate solution. All values outside of the active region have zero fitness contribution. The fitness contributions of each position (i.e., each trait) define the "phenotype" of the candidate solution; two candidate solutions that differ only in inactive regions will have identical phenotypes. Figure 5.1 shows an example fitness calculation. Given this search space, the optimal solution will have a 100.0 in every position of its vector starting from the very first, making the entire candidate solution active and each value maximized. However, any candidate solution with an activation position other than the first will not have a pathway to the global optimum that is reachable via hill climbing alone.

Given the large number of pathways that need to be simultaneously explored, this diagnostic allows us to compare the exploration capacity of different selection schemes. Additionally, this diagnostic allows researchers to test the exploration breaking point of a given selection scheme, as increasing the cardinality of the diagnostic increases the exploratory capacity needed to find the best activation position. In this work, we use this diagnostic to test the exploratory limits of lexicase selection along with a number of its variants.

5.3 Lexicase Selection

Reference [36] introduced the lexicase parent selection algorithm for solving GP problems that require programs to produce qualitatively different modes of response for different inputs. Since its introduction, lexicase selection has been demonstrated to be successful across a broad range of problem domains, including automatic program synthesis [15], symbolic regression [25], evolutionary robotics [31], genetic algorithms [29], and learning classifier systems [1].

In lexicase selection, individuals are evaluated on a set of selection criteria (e.g., test cases or other types of fitness functions). For each selection event, each member of the population is initially considered to be a candidate for selection. To select an individual, lexicase shuffles the set of test cases, and considers each test case in sequence. In shuffled order, each test case is used to filter the candidates, removing all but the best individuals from further consideration. This process of winnowing candidates continues until only one candidate remains to be selected or until all test cases have been considered; if more than one candidate remains, one is selected at random. Algorithm 5.1 details the lexicase selection algorithm.

Algorithm 5.1 Lexicase selection for a single parent. Adapted from [18].

1. Mark entire population as current **candidates** under consideration.
2. Shuffle **test_cases** into a random order.
3. For each **case** in **test_cases**:
   a. Evaluate each candidate in **candidates** on **case**.
   b. Identify the **best_score** on **case** of all candidates.
   c. Remove each entry from **candidates** with a score on **case** worse than **best_score**.
4. Select a random entry from **candidates**.

Because the set of test cases are shuffled whenever a parent must be chosen, individuals that perform well on different partitions of the test set can coexist within the population [5]. Indeed, this dynamic creates niches where different members of the population can specialize on different subsets of selection criteria, allowing a population to simultaneously explore many pathways to solving a given problem. Moreover, this focus on exploration does not necessarily sacrifice lexicase's ability to exploit each pathway since only the best performing individuals are selected for a given sequence of test cases.

Many variants of lexicase selection have been proposed, each either specialized for solving a particular type of problem or designed to address potential short comings of the original lexicase selection scheme. Below, we describe each of the four variants of lexicase selection examined in this work.

5.3.1 Epsilon Lexicase Selection

Epsilon lexicase selection relaxes the elitism of the filtering step in standard lexicase selection (step 3c in Algorithm 5.1). When filtering candidates on a given test case, epsilon lexicase retains all individuals with performances within some threshold (ε) of the best performance on that test case. The ε parameter can be tuned by the practitioner and can be applied either as a proportion of the optimal performance on a given test case or as an absolute threshold.

Epsilon lexicase selection specializes standard lexicase selection for problems where performances on selection criteria are measured using real-valued numbers, such as symbolic regression problems [25, 34, 37] or evolving robot controllers [30, 31]. The standard lexicase selection algorithm assumes that individuals with equivalent performances on a given test case will have equal scores for that test case. Inconsequential noise in an individual's score on a particular test case could result in arbitrary, but consequential differences in which individuals are selected by the standard lexicase algorithm. By allowing a small ε difference between individuals, epsilon lexicase addresses this potential problem.

In this work, we vary ε to investigate how it affects exploration. Reference [25] observed that behavioral diversity increases at larger values of ε. Given ε's affect on behavioral diversity, we hypothesize that increasing ε will increase the exploration capacity of epsilon lexicase. However, at too high of an ε value, we expect *meaningful* exploration to degrade. That is, as ε increases beyond a certain point, different adaptive pathways blur together as meaningful differences in test case performances become indistinguishable.

For simplicity, we apply ε as a fixed absolute error threshold in this work. Future work, however, should investigate how different applications of ε further influence lexicase's exploration capacity (e.g., semi-dynamic and dynamic applications of ε from [24]).

5.3.2 Down-Sampled Lexicase Selection

Down-sampled lexicase applies random subsampling to the selection criteria in order to reduce the per-generation computational effort required by lexicase selection [7, 19]. Down-sampled lexicase uses a random subset of test cases each generation, which reduces the number of test cases on which each individual in the population must be evaluated every generation. After down sampling, the standard lexicase procedure is used to choose parents.

For an equivalent number of total evaluations, down-sampled lexicase allows practitioners to run their evolutionary computing system for more generations or with a larger population size; both of which have been shown to improve problem-solving success [7, 16, 19]. In this work, we investigate how down sampling affects lexicase selection's exploratory capacity. While [7] found no evidence that down sampling reduces phenotypic diversity across a range of program synthesis problems, they did find that down sampling degrades specialist maintenance. We hypothesize that down sampling's negative effect on specialist maintenance harms its exploratory capacity. Entire categories of test cases may be excluded on any given generation, and candidate solutions specializing on those test cases may be lost as a result. Such dynamics may prevent extensive exploration of valuable niches.

5.3.3 Cohort Lexicase Selection

Cohort lexicase partitions the test case set and the population each into an equal number of cohorts. Each generation, cohort membership is randomly assigned, and each cohort of candidate solutions is paired with a cohort of test cases. Each cohort of candidate solutions is evaluated only on the test cases in the paired test case cohort, which, like down-sampled lexicase, reduces the required number of per-generation evaluations relative to standard lexicase selection. Unlike down-sampled lexicase, however, cohort lexicase ensures that every test case in the full set is used every generation, as each cohort of candidate solutions competes on a different subset of the full set. To select a parent, cohort lexicase first selects a cohort to choose from; previous work guaranteed an equal number of parents were selected from each cohort each generation [7, 19]. Candidate solutions only compete against other solutions within their respective cohort, and within-cohort competition is arbitrated by the test cases in the associated cohort of tests.

In this work, we investigate how the number of cohorts that we partition the population and test set into influences lexicase selection's capacity for exploration. For similar reasons to down-sampled lexicase, we expect cohort lexicase selection to degrade lexicase selection's exploratory capacity. However, because cohort lexicase uses every test case in every generation, we expect it to better support exploration than down-sampled lexicase. As we increase the size of cohorts (and decrease the number of cohorts), we expect cohort lexicase to approach the exploratory abilities of standard lexicase selection. This could be due to the fact that as cohort size increases, the chances of a specialist being paired with the test cases it specializes on also increases.

5.3.4 Novelty-Lexicase Selection

Novelty-lexicase selection combines standard lexicase selection with novelty search [22]. Novelty search disregards functional objectives and instead searches for behavioral novelty, steering populations to continuously explore new regions of the search space [28]. As such, novelty search is argued to be well-suited for solving problems with deceptive fitness landscapes where local gradients lead *away* from the global optimum [27]. Novelty-lexicase selection incorporates ideas from novelty search into lexicase selection.

Novelty-lexicase selection (as introduced in [22]) requires that the entire population be evaluated on all test cases. For each member of the population, novelty-lexicase selection computes their "novelty score" on each test case. A novelty score measures how different a candidate solution's output on a given test case is from the rest of the population. Here, a candidate solution's novelty score on a test case equals the average distance between its output and the k nearest neighbor outputs for that test case. Novelty-lexicase selection incorporates novelty scores by augmenting the test case set with an additional novelty test case for every original test case. Using this augmented set of test cases, the standard lexicase procedure is used to choose parents.

In this work, we use our exploration diagnostic to compare the exploratory capacity of novelty-lexicase selection (at k =1, 2, 4, 8, 15, 30, and 60) and standard lexicase selection ($k = 0$). Reference [22] found that novelty-lexicase selection generally maintained more behavioral diversity than standard lexicase selection on several program synthesis problems. As such, we expect the addition of novelty score test cases to improve lexicase selection's exploratory capacity on our exploration diagnostic.

5.4 Diagnosing the Exploratory Capacity of Lexicase Selection and Its Variants

We conducted a series of experiments to analyze the exploratory limits of standard lexicase selection and four of its variants: epsilon lexicase, down-sampled lexicase, cohort lexicase, and novelty-lexicase. For each experiment, unless stated otherwise, we evolved populations of 500 numerical vectors on our exploration diagnostic with a cardinality of 100 for 50,000 generations. Across all experiments, we ran 50 replicates of each constituent treatment. We initialized populations to the lowest point in the fitness landscape, vectors of all 0.0 s.

When evaluating a candidate solution, we calculated a score associated with each position in its vector according to the exploration diagnostic (Fig. 5.1). We used this collection of scores as test case qualities for lexicase selection and its variants. For this work, we report quality directly; for comparison to other studies, note that test case *error* is the amount that quality is below 100. When a single fitness value was required (e.g., for tournament selection), we summed the individual test case qualities to determine the solution's aggregate fitness.

Selected candidate solutions reproduced asexually, and we applied point-mutations to offspring at a per-position rate of 0.7%. The magnitude of each mutation was drawn from a normal distribution with a mean of 0.0 and a standard deviation of 1.0 ($\mathcal{N}(0, 1)$). When mutations would raise a trait to a value x where $x > 100$, we rebounded that trait to $200 - x$, ensuring that each trait value remained less than or equal to 100. When mutations would lower a trait below 0.0, we reset that trait to 0.0.

For each replicate of each experiment, we extracted the most performant individual in the population (i.e., the individual with the highest aggregate score) to compare across treatments. For different diagnostic cardinalities (i.e., different numbers of test cases), the range of possible aggregate scores differs; as such, we normalized all aggregate scores by dividing by the cardinality, which results in a value between 0.0 and 100.0.

To identify the number of pathways being explored by a population, we measured the number of unique activation positions within each population. Using this measurement, we calculated "activation position coverage" as the fraction of possible activation positions represented in a population.

For each experiment, we report both mean performance and mean activation position coverage over time (each with a bootstrapped 95% confidence interval), and we compare measurements from the final generation across treatments. For each comparison, we performed a Kruskal–Wallis test to determine if there were significant differences; if so, we applied a Wilcoxon rank-sum test to distinguish between pairs of treatments, applying Bonferroni corrections for multiple comparisons where appropriate.

The software used to conduct experiments, statistical analyses, experimental data, and guides for replication are included in our supplemental material [20]. See Sect. 5.6 for more details.

5.4.1 Lexicase Selection Out-Explores Tournament Selection

First, we used the exploration diagnostic to test well-established expectations that lexicase selection improves search space exploration relative to tournament selection. Unlike lexicase selection, tournament selection does not reliably maintain multiple niches within a population [5]; as such, we expected it to perform worse than lexicase selection on the exploration diagnostic. For this experiment, we used tournaments of eight individuals.

Consistent with our expectations, we found that lexicase selection outperforms tournament selection on the exploration diagnostic (Fig. 5.2; Wilcoxon rank-sum test: $p < 10^{-4}$). Early on, populations evolving under tournament selection converge to a single local optimum in the exploration diagnostic (i.e., a single activation position); without a mechanism to escape, populations become stuck and fail to continue exploring the search space. Lexicase selection, however, rewards specialists for different activation positions, allowing the population to continuously explore different evo-

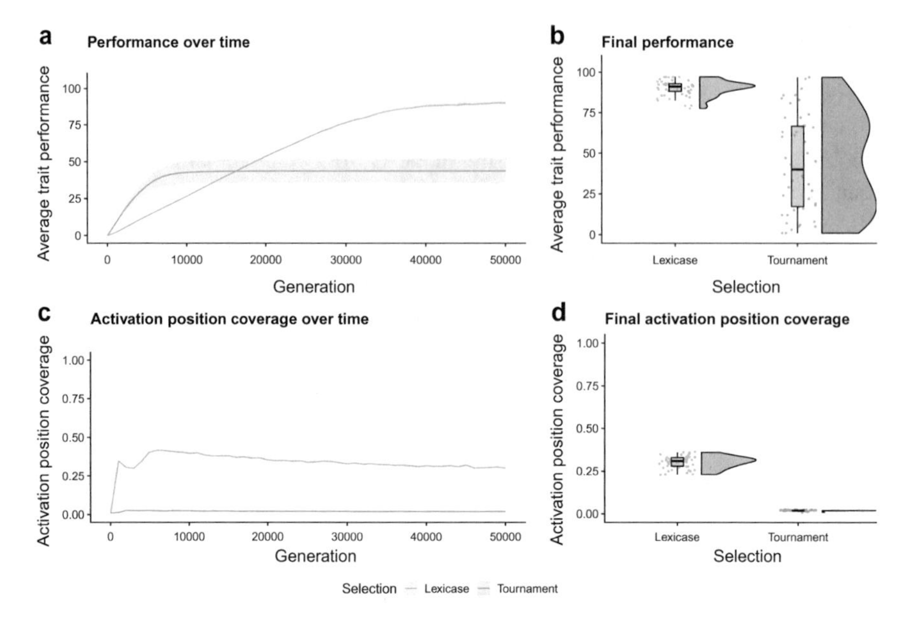

Fig. 5.2 Lexicase selection versus tournament selection on the exploration diagnostic. Panels **a** and **b** show performance over time and at the end of 50,000 generations, respectively. Likewise, panels **c** and **d** show activation position coverage over time and at the end of 50,000 generations, respectively. For panels **a** and **c**, each line gives the mean value across 50 replicates, and the shading around each mean gives a 95% confidence interval

lutionary pathways. Indeed, we found that lexicase selection maintains substantially more "activation-position" specialists than tournament selection (Fig. 5.2; Wilcoxon rank-sum test: $p < 10^{-4}$).

5.4.2 *The Exploratory Capacity of Lexicase Selection Degrades as We Increase Diagnostic Cardinality*

Next, we evaluated standard lexicase selection on the exploration diagnostic at cardinalities 10, 20, 50, 100, 500, and 1,000. Cardinality defines the number of potential pathways that must be explored by a population to guarantee finding the global optimum; increasing cardinality obscures the path to optimality. Cardinality also corresponds to the number of test cases (i.e., niches) that individuals can specialize on. For a fixed population size, increasing the number of test cases decreases the long-term survival probability of any single specialist under lexicase selection [5], which could negatively affect lexicase's capacity to fully explore pathways in the search space. For these reasons, we expected lexicase selection's performance on the exploration diagnostic to degrade as we increased cardinality.

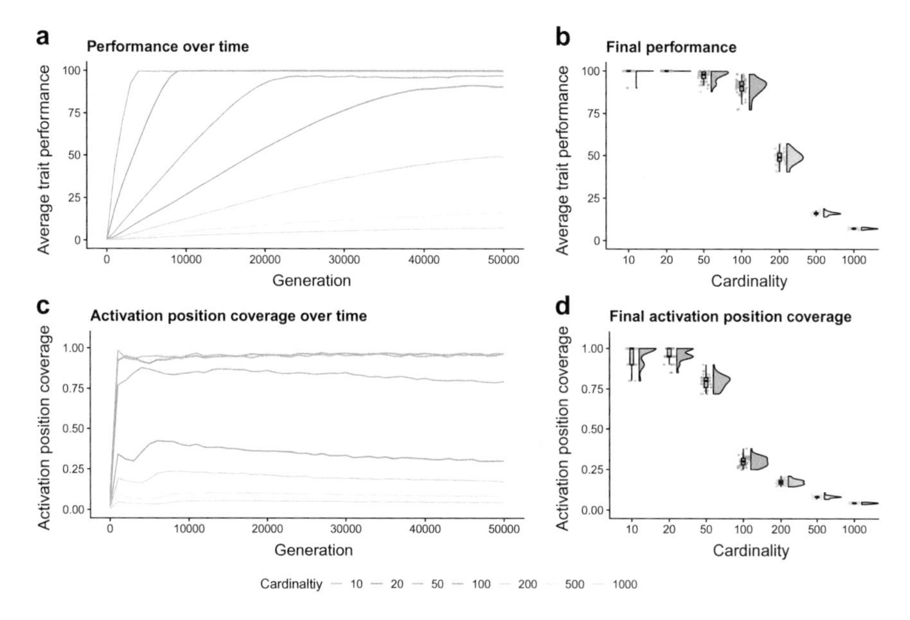

Fig. 5.3 Lexicase selection at a range of exploration diagnostic cardinalities. Panels **a** and **b** show performance over time and at the end of 50,000 generations, respectively. Likewise, panels **c** and **d** show activation position coverage over time and at the end of 50,000 generations, respectively. For panels **a** and **c**, each line gives the mean value across 50 replicates, and the shading around each mean gives a 95% confidence interval

Figure 5.3 shows lexicase selection's performance at each cardinality of the exploration diagnostic. Across all cardinalities, lexicase selection improves performance over time. Notably, treatments with cardinalities 10, 20, and 50 each perform near optimally after 50,000 generations, and populations evolved under cardinality 100 perform relatively well. Higher cardinalities (e.g., 200, 500, and 1000), however, perform substantially worse (Wilcoxon rank-sum tests: $p < 10^{-4}$) and appear to need more time to converge on their maximal performance. These data verify that increasing diagnostic cardinality also increases the exploration diagnostic's difficulty, as lexicase selection's performance degrades as cardinality increases.

We also found that populations evolved at lower diagnostic cardinalities maintained a larger coverage of unique activation positions than populations evolved at higher diagnostic cardinalities (Fig. 5.3). Such diversity maintenance likely drove lexicase selection's ability to continuously explore pathways in the search space.

In these experiments, we used a population size of 500, resulting in 500 selection events per generation. In each selection event, scores for vector positions (Fig. 5.1) are prioritized in a random order. Across a population, we expect that positions that are consistently rewarded should maintain solutions that start at that position. The optimal solution requires the initial position to be the highest in the population, but this position may, by chance, never be evaluated first during lexicase selection. The probability of this occurring varies with cardinality. With a population size of

500 and a vector with 50 positions (i.e., a diagnostic cardinality of 50), there is a 0.004% chance (1 in 25,000) of the initial position never being chosen first in a generation, making it unlikely to go unselected. Increasing the cardinality to 100, however, increases the chance for the first position to go unselected to 0.657% (1 in 152)—a much more likely occurrence that may explain the reduced performance at cardinality 100 relative to cardinality 50. By cardinality 200, the probability for the first position to go unselected within a given generation rises to 8.157%, an even more likely occurrence.

One way to combat these dynamics is to increase population size, which would allow lexicase selection to support higher levels of exploration by reducing the chances of any given starting position from being skipped over by selection in any single generation. However, increasing population size can be computationally expensive, as more individuals would need to be evaluated every generation. Decreasing the depth of evolutionary search by reducing the number of generations evaluated is one way to balance the cost of increasing population size. For a fixed computational budget, can increasing population size at the expense of evaluating fewer generations of evolution pay off under lexicase selection?

5.4.3 Increasing Population Size Can Improve Lexicase Selection's Exploratory Capacity

To test whether increasing population size can improve lexicase selection's exploratory capacity, we extended the runtime of our experiment and compared lexicase selection's performance on the exploration diagnostic (with a cardinality of 100) at two population sizes: 500 and 1,000. Because increasing population size increases per-generation computational effort, we ran both conditions for a fixed number of test case evaluations, evolving populations of 500 individuals for twice as many generations as populations of 1,000 individuals (1,000,000 and 500,000 generations, respectively). As such, lineages from 500-individual populations take two reproductive steps in the search space for every one step reproductive step taken by a 1000-individual population. This difference may allow the smaller populations to more rapidly exploit their initial position in the search space. However, if larger populations are able to maintain more pathways in the search space, they may eventually outperform smaller populations.

As expected, we found that increasing population size allows lexicase selection to maintain more starting positions for the entire duration of our experiment (Fig. 5.4). Smaller populations initially outperform larger populations (given a fixed computational budget); however, despite running for fewer total generations, larger populations eventually outperform the smaller populations (Fig. 5.4; Wilcoxon rank-sum test: $p < 10^{-4}$). These data suggest that, for a fixed number of test case evaluations, we can indirectly tune lexicase selection's level of search space exploitation and exploration by adjusting our allocation of computational resources between generations of evolution and population size.

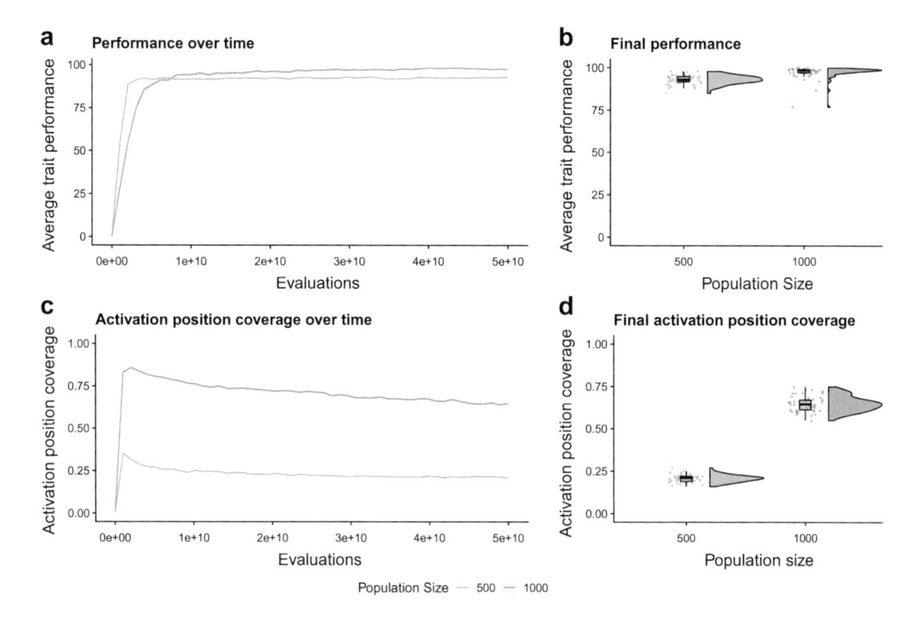

Fig. 5.4 Lexicase selection's performance on the exploration diagnostic at different population sizes. Panels **a** and **b** show performance over time and at the end of the experiment, respectively. Likewise, panels **c** and **d** show activation position coverage over time and at the end of the experiment, respectively. For panels **a** and **c**, each line gives the mean value across 50 replicates, and the shading around each mean gives a 95% confidence interval

5.4.4 *Relaxing Lexicase Selection's Elitism Can Improve Exploration*

As discussed in Sect. 5.3.1, epsilon lexicase relaxes the elitism of lexicase selection. To test whether this relaxation of elitism affects exploration, we compared standard lexicase selection and epsilon lexicase selection on the exploration diagnostic. Specifically, we evolved 50 replicate populations at each of the following ε values: 0.0 (standard lexicase), 0.1, 0.3, 0.6, 1.2, 2.5, 5.0, and 10.0.

Epsilon lexicase with small values of ε (0.1 and 0.3) outperforms standard lexicase selection on the exploration diagnostic (Fig. 5.5; Wilcoxon rank-sum tests: $p < 10^{-4}$). Extreme values of ε (5.0 and 10.0) significantly degrade performance relative to standard lexicase selection (Wilcoxon rank-sum tests: $p < 10^{-4}$). Interestingly, intermediate values of ε (0.6 and 1.2) perform best during the first approximately 20,000 generations, but are eventually outperformed by treatments with smaller values of ε. Unlike previous experiments, the relative levels of activation position coverage among conditions does not correspond with diagnostic performance.

In general, epsilon lexicase is expected to have two main advantages over standard lexicase selection [25]: (1) it allows small amounts of noise in the evaluation data to be

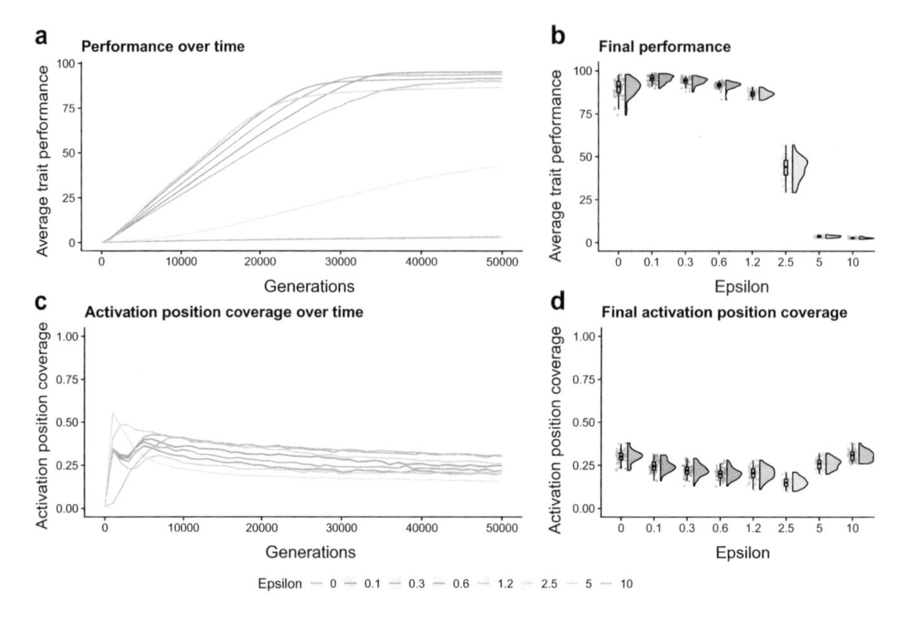

Fig. 5.5 Epsilon lexicase selection's performance on the exploration diagnostic at a range of ε values. Panels **a** and **b** show performance over time and after 50,000 generations of evolution, respectively. Likewise, panels **c** and **d** show activation position coverage over time and after 50,000 generations of evolution, respectively. For panels **a** and **c**, each line gives the mean value across 50 replicates, and the shading around each mean gives a 95% confidence interval

ignored, and (2) it prevents nearly identical scores from determining which candidate solutions win, potentially allowing for greater coexistence. While the first mechanism cannot be at play here (since all scores are deterministic), the second advantage could provide additional support for solutions further along a given pathway. That is, solutions that begin optimizing at an earlier point in their vector, by definition, must have slightly lower values for later positions in their activated region. In standard lexicase, when two solutions had overlapping activation regions, the one that start later would have an advantage at all overlapped sites. In epsilon lexicase, however, the earlier start (i.e., the one with more long-term potential) now has a better chance to pass lexicase selection's selective filter.

5.4.5 Down-Sampling Degrades Lexicase Selection's Exploratory Capacity

Next, we investigated whether down-sampling affects lexicase selection's exploratory capacity by comparing the performance of lexicase selection at a range of sampling rates: 100% (standard lexicase), 50%, 20%, 10%, 5%, 2%, and 1%. For example,

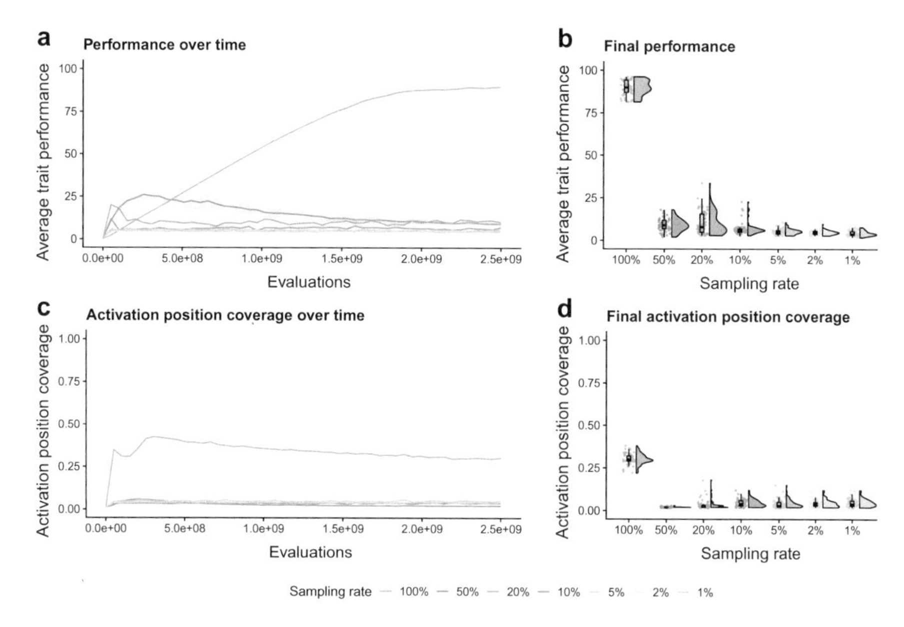

Fig. 5.6 Down-sampled lexicase selection's performance on the exploration diagnostic at a range of subsampling rates. Panels **a** and **b** show performance over time and at the end of the experiment, respectively. Likewise, panels **c** and **d** show activation position coverage over time and at the end of the experiment, respectively. For panels **a** and **c**, each line gives the mean value across 50 replicates, and the shading around each mean gives a 95% confidence interval

a 10% sampling rate means that each generation we randomly selected 10 of the 100 possible test cases (for a diagnostic cardinality of 100) to be used for parent selection. Down-sampling reduces the per-generation computational effort required for parent selection by conducting fewer test case evaluations (Sect. 5.3.2). For a fair comparison across different sampling rates, we limited the computational budget to a maximum of 2.5×10^9 test case *evaluations* by varying the number of generations of evolution for each subsampling rate (100%: 50,000 generations, 50%: 100,000 generations, 20%: 250,000 generations, 10%: 500,000 generations, 5%: 1,000,000 generations, 2%: 2,500,000 generations, and 1%: 5,000,000 generations).

Any amount of down-sampling significantly degraded lexicase selection's performance on the exploration diagnostic for the allotted computational budget (Fig. 5.6; Wilcoxon-rank sum tests: $p < 10^{-4}$). Down-sampled lexicase selection's drop in performance is likely attributed to frequent mismatches between candidate solutions and the test cases that they are specialized on. As the proportion of test cases used in each generation decreases, so too does the probability of a solution encountering the same set of test cases for multiple generations in a row. As such, a solution has a reduced chance of encountering the test cases for which it is most optimized [7]. These dynamics will repeatedly remove solutions with small active regions, thereby reducing population diversity. Indeed, we found that down-sampling substantially reduces the number of activation position specialists represented in the population (Fig. 5.6; Wilcoxon rank-sum tests: $p < 10^{-4}$). In fact, any down-sampling used appears to have a strong negative effect, substantially reducing performance in all cases.

We repeated this experiment, except we increased population size instead of increasing generations of evolution for down-sampled lexicase; that is, we ran each condition for an equivalent number of generations but differing population sizes to maintain a fixed number of evaluations. We report these data in our supplemental material [20]. Overall, the patterns were similar to that of increasing generations of evolution. Initially, down-sampled lexicase outperforms standard lexicase on the exploration diagnostic; however, standard lexicase eventually outperforms down-sampled lexicase across all subsampling rates [20].

5.4.6 Cohort Partitioning Degrades Lexicase Selection's Exploratory Capacity

Next, we evaluated whether partitioning the population and test cases into cohorts affects the exploration capacity of lexicase selection. We compared the performance of standard lexicase to that of cohort lexicase at a range of cohort sizes (given as the proportion of the population and the set of test cases used in each cohort): 100% (standard lexicase), 50%, 20%, 10%, 5%, 2%, and 1%. For example, a cohort size of 10% means that the population (of 500 individuals) is divided into 10 cohorts of 50 individuals each, and the test cases (100 total) are also divided into those same 10 cohorts, with 10 test cases in each. Like down-sampled lexicase, cohort lexicase reduces the per-generation computational effort required for parent selection by evaluating each cohort of candidate solutions on only one of the test case cohorts (Sect. 5.3.3). Likewise, for fair comparison across different cohort sizes, we limited the computational budget to a maximum of 2.5×10^9 test case evaluations by varying the number of generations of evolution for each cohort size (100%: 50,000 generations, 50%: 100,000 generations, 20%: 250,000 generations, 10%: 500,000 generations, 5%: 1,000,000 generations, 2%: 2,500,000 generations, and 1%: 5,000,000 generations).

As with down-sampled lexicase, any level of cohort partitioning degrades lexicase's performance on the exploration diagnostic for the allotted computational budget (Fig. 5.7; Wilcoxon rank-sum tests: $p < 10^{-4}$). However, cohort lexicase does not appear to degrade lexicase selection's performance to the same degree as down-sampled lexicase for a given subsampling rate (Fig. 5.6). Moreover, standard lexicase took longer (more total evaluations) to outperform cohort lexicase than to outperform down-sampled lexicase. These data suggest that cohort partitioning (with intermediate levels of partitioning) may be a better method of random subsampling in the context of lexicase selection.

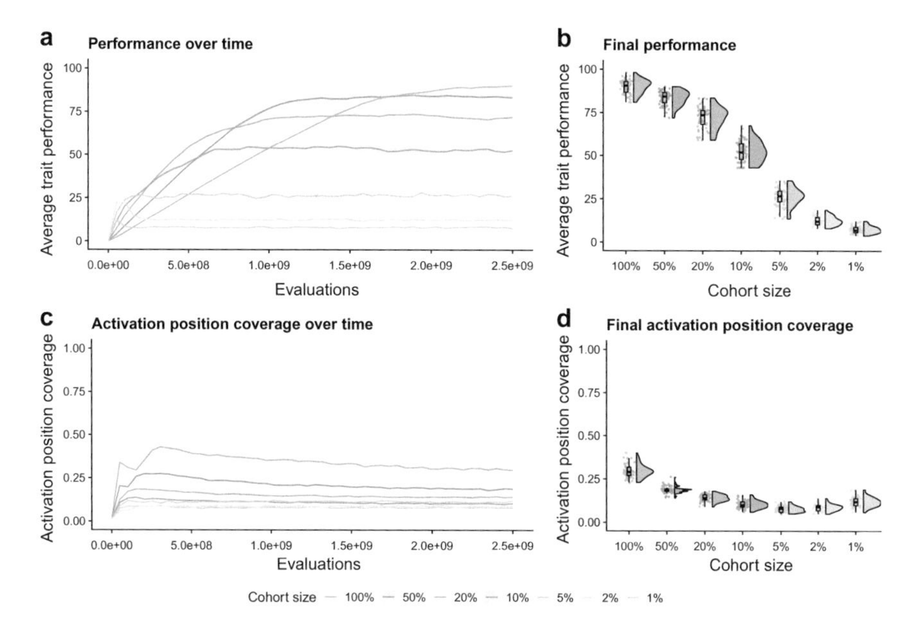

Fig. 5.7 Cohort lexicase selection's performance on the exploration diagnostic at a range of partitioning rates. Panels **a** and **b** show performance over time and at the end of the experiment, respectively. Likewise, panels **c** and **d** show activation position coverage over time and at the end of the experiment, respectively. For panels **a** and **c**, each line gives the mean value across 50 replicates, and the shading around each mean gives a 95% confidence interval

We repeated this experiment, except we increased population size instead of increasing generations of evolution for cohort lexicase; that is, we ran each condition for an equivalent number of generations but differing population sizes to maintain a fixed number of evaluations. We report these data in our supplemental material [20]. The overall patterns were qualitatively different and warrant further exploration in future work. We found no compelling evidence that cohort lexicase outperformed standard lexicase in the given computational budget; however, we did find that populations evolving under cohort lexicase (with larger population sizes) maintained more activation position coverage than standard lexicase selection [20]. Further, some of the cohort sizes were on an upward trajectory when the runs finished and may eventually outperform standard lexicase given a larger computational budget.

5.4.7 *Cohort Lexicase Out-Explores Down-Sampled Lexicase*

Next, we independently verified that cohort lexicase out-explores down-sampled lexicase on the exploration diagnostic. To do so, we compared the performance of cohort lexicase and down-sampled lexicase with their most performant parameteri-

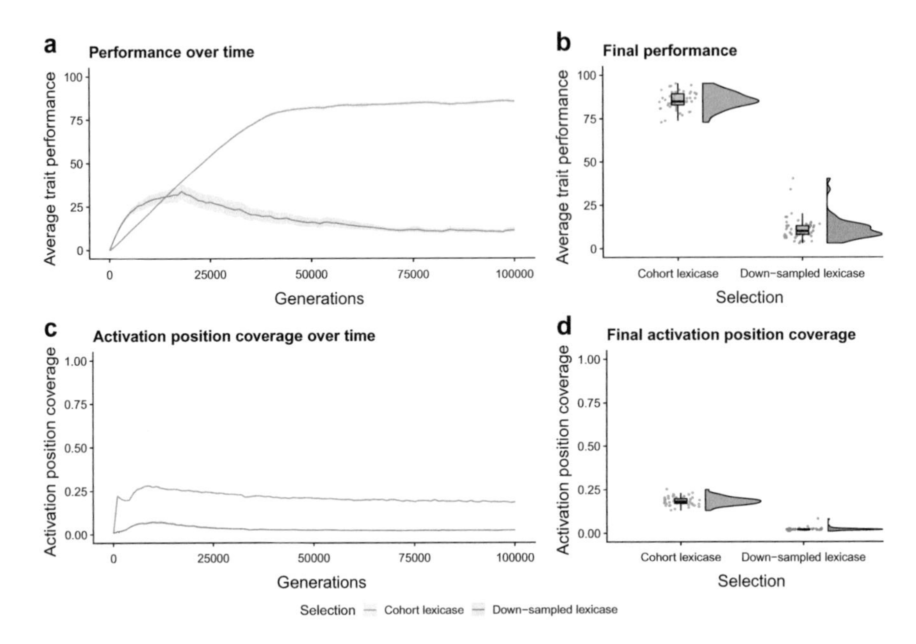

Fig. 5.8 Down-sampled versus cohort lexicase on the exploration diagnostic. Panels **a** and **b** show performance over time and at the end of the experiment, respectively. Likewise, panels **c** and **d** show activation position coverage over time and at the end of the experiment, respectively. For panels **a** and **c**, each line gives the mean value across 50 replicates, and the shading around each mean gives a 95% confidence interval

zations: a 50% cohort size and a 50% sampling rate, respectively. We again limited the computational budget to a maximum of 2.5×10^9 test case evaluations (100,000 generations of evolution for both conditions), and we ran 50 new replicates of each condition for comparison.

As expected given Figs. 5.6 and 5.7, cohort lexicase outperformed down-sampled lexicase by a substantial margin for the given computational budget (Fig. 5.8; Wilcoxon rank-sum test: $p < 10^{-4}$). Interestingly, down-sampled lexicase appears to briefly outperform cohort lexicase in the first few thousand generations but is quickly overtaken by cohort lexicase. Both cohort and down-sampled lexicase offer equivalent per-generation evaluation savings, but cohort lexicase uses every test case for parent selection in every generation. This could play a role in problem-solving success, as a test case that rewards exploration at any given activation position in the exploration diagnostic is used every generation. Indeed, populations evolving under cohort lexicase selection maintained a higher diversity of activation positions than populations evolving under down-sampled lexicase selection (Fig. 5.8; Wilcoxon rank-sum test: $p < 10^{-4}$).

Previous work predicted the potential for such differences between cohort and down-sampled lexicase. Reference [7] found that cohort lexicase better maintained phylogenetic diversity than down-sampled lexicase, as phylogenies coalesced less frequently under cohort lexicase selection (maintaining deeper, more divergent branches). Despite this difference in diversity maintenance, [7] did not find significant differences in problem-solving success across a set of program synthesis benchmark problems, which suggests that the test cases used in these benchmark problems were more robust to random subsampling than the test cases for the exploration diagnostic. Indeed, each individual test case for the exploration diagnostic uniquely represents a single activation position; that is, test cases are minimally redundant with one another. In many program synthesis benchmark problems, however, individual test cases are often intentionally redundant to others, differing only in the particular values of their inputs and outputs and not necessarily different in the functional specialization they reward. Such redundancies prevent candidate solutions from memorizing particular input-output pairings, forcing candidate solutions to generalize in order to achieve high fitness across redundant test cases. This detail could explain why the exploration diagnostic reveals substantial performance differences between cohort and down-sampled lexicase where more standard benchmark problems failed to do so.

5.4.8 Novelty Test Cases Degrade Lexicase Selection's Exploratory Capacity

Finally, we evaluated how incorporating novelty test cases into lexicase selection impacts exploration. We compared the performance of standard lexicase to that of novelty-lexicase for a range of k-nearest neighbors: 0 (standard lexicase), 1, 2, 4, 8, 15, 30, and 60.

Contrary to our expectations, we found that the addition of novelty test cases degrades performance on the exploration diagnostic in all cases (Fig. 5.9; Wilcoxon rank-sum test: $p < 10^{-4}$). Though, novelty-lexicase generally maintains similar levels of activation position diversity in the population relative to standard lexicase, and by the end of the experiment, some parameterizations of novelty lexicase maintain more activation positions, though none of the differences appear to be substantial (Fig. 5.9).

Novelty search favors solutions that have never been seen before, regardless of their impact on fitness. Based on previous studies, we expected novelty-lexicase to outperform standard lexicase on the exploration diagnostic [22]. However, novelty-lexicase appears to hinder lexicase's ability to fully exploit pathways in the diagnostic's search space.

While past work has demonstrated that novelty search can be effective at producing solutions for complicated problems, the exploration diagnostic does not have any of the hidden intricacies that novelty search excels at disentangling. Indeed, novelty search appears to thrive under conditions where there are more non-linearities between genotype and phenotype. The underlying representation used here is purposely simple numerical vectors as opposed to an artificial neural network [27] or

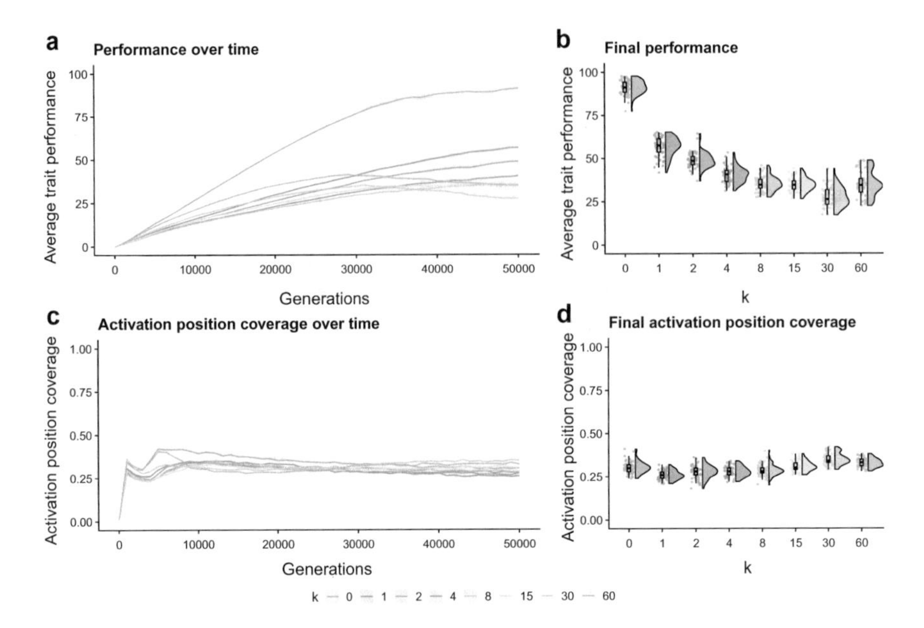

Fig. 5.9 Novelty-lexicase selection's performance on the exploration diagnostic at a range of nearest-neighbor parameterizations. Panels **a** and **b** show performance over time and after 50,000 generations of evolution, respectively. Likewise, panels **c** and **d** show activation position coverage over time and after 50,000 generations of evolution, respectively. For panels **a** and **c**, each line gives the mean value across 50 replicates, and the shading around each mean gives a 95% confidence interval

PushGP [22] where internal architectures can change and qualitatively different outputs are possible. For example, in this case, all sites in a genome are optimal at one end of their range of values, whereas most complex problems are assumed to have pockets of solutions throughout the genotype-phenotype map. Additionally, our results also used a single, limited form of novelty lexicase. We did not use a seed bank (the importance of which has previously been stressed), and we used k-nearest neighbors euclidean distances to measure novelty instead of a direct measure of behavioral uniqueness. These differences in problems may shine a light as to why novelty-lexicase did not outperform standard lexicase selection on the exploration diagnostic.

Our results from varying diagnostic cardinality (Sect. 5.4.2) may also offer insights into the unexpectedly poor performance of novelty-lexicase selection. Novelty-lexicase selection increases the number of test cases used for parent selection (in this work, doubling the number of test cases from 100 to 200). Increasing the number of test cases (without simultaneously increasing the population size) is not without cost, degrading specialist maintenance and performance on the exploration diagnostic (Fig. 5.3). This dynamic is likely to be at play in our novelty-lexicase experiment, as population size was constant for both standard lexicase and novelty-lexicase selection.

5.5 Conclusion

In this work, we introduced a new diagnostic to investigate the exploratory limits of lexicase selection along with several of its variants: epsilon lexicase, down-sampled lexicase, cohort lexicase, and novelty-lexicase. First, we verified well-established expectations that lexicase selection better facilitates search space exploration than tournament selection. Across all exploration diagnostic difficulty levels (i.e., cardinalities), lexicase selection drove improvements in performance (Fig. 5.3), while tournament selection repeatedly failed to escape early local optima (Fig. 5.2). As we increased the cardinality of the diagnostic, lexicase selection's specialist maintenance and overall performance waned. Conditions with larger diagnostic cardinalities used more test cases to evaluate individuals, and as such had more possible specialists (i.e., niches). Given a fixed population size, lexicase maintained a smaller fraction of possible specialists as the number of possible niches increased, which, in turn, decreased overall performance (Fig. 5.3).

Interestingly, we found that allocating a computational budget (i.e., candidate solution evaluations) toward increasing generations versus increasing population size is not necessarily a straightforward choice when using lexicase selection. In our case, a larger population size enabled better specialist maintenance and ultimately higher performance on the exploration diagnostic with standard lexicase (Fig. 5.4). This finding is interesting in light of [17]'s work investigating the problem-solving benefits of down-sampled lexicase; on a suite of program synthesis problems, Helmuth and Spector found that some problems benefited from an increased population size (at the cost of running for fewer generations), some problems benefited from an increase in generations, and most problems were unaffected by their choice of increasing population size versus generations evaluated.

Overall, these results suggest that lexicase selection can be sensitive to expanding the set of test cases used for evaluation, especially if each test case uniquely represents a distinct, desirable trait. Moreover, our results suggest the importance of more deeply examining the benchmark problems that we use and the characteristics of the search spaces that they represent. Given a fixed computational budget, why do some problems benefit from running deeper evolutionary searches while others benefit from increased population sizes under lexicase selection? For many problems, different categories of test cases have uneven representation in the test set. We hypothesize that the distribution of test cases among categories plays a role in lexicase selection's success and the optimal balance between population size and depth of search (generations of evolution). For example, if the number of test cases is similar to population size, lexicase selection may fail to maintain specialists on categories that are underrepresented in the test cases and instead favor overrepresented categories. In future work, we will develop novel diagnostic tools for investigating the sensitivity of selection schemes to test case set composition.

We found that each of the lexicase variants that we evaluated—epsilon lexicase, down-sampled lexicase, cohort lexicase, and novelty-lexicase—affected lexicase selection's exploratory capacity. For small values of ε, epsilon lexicase outperformed standard lexicase selection on the exploration diagnostic, while large values of ε substantially degraded performance. Surprisingly, we found that novelty-lexicase degrades performance on the exploration diagnostic relative to standard lexicase selection.

Our experiments are also the first to demonstrate consequential differences between down-sampled and cohort lexicase selection, as previous work generally failed to distinguish the problem-solving performance of these two lexicase variants [7]. Cohort lexicase substantially outperformed down-sampled lexicase (Fig. 5.8). Both down-sampled and cohort lexicase offer equivalent per-generation evaluation savings, so our results suggest that cohort partitioning may often be a better subsampling method than down-sampling for lexicase selection. Future work should examine whether this difference between cohort partitioning and down-sampling holds across different selection schemes.

Given equivalent computational budgets, we found that standard lexicase selection eventually outperforms both cohort and down-sampled lexicase on the exploration diagnostic (Figs. 5.6 and 5.7). This result diverges from recent benchmarking studies where subsampling substantially improved performance on a range of program synthesis problems [7, 16, 17]. Future work will develop diagnostic problems to help identify when subsampling (e.g., via either cohort partitioning or down-sampling) is likely to improve versus impede lexicase selection's performance.

In each of our experiments, we focused our analyses on performance and activation position diversity maintenance. Future work should more deeply examine the evolutionary histories of evolving populations using phylodiversity metrics [4]. Along with this, other parameter values and configurations of each of the variants evaluated in this work could be tested in order to develop a more complete understanding of how parameterization affects exploration.

We intend for this work to demonstrate how diagnostics (e.g., the exploration diagnostic introduced here) can be valuable tools for evaluating the pros and cons of different selection schemes. We plan to implement a larger suite of selection scheme diagnostics, each targeted toward evaluating a particular aspect of problem-solving. Such diagnostics will complement conventional benchmarking experiments in our community's effort to understand how different selection schemes steer evolutionary search.

5.6 Data and Software Availability

Our supplemental material [20] is hosted on GitHub and contains the software, data analyses, and documentation associated with this work. Our experiments are implemented using the Empirical library [33], and we used a combination of Python and R version 4 [35] for data processing and analysis. We used the following R packages for data wrangling, statistical analysis, graphing, and visualization: ggplot2 [39], tidyverse [38], knitr [42], cowplot [40], viridis [8], RColorBrewer [32], rstatix

[23], ggsignif [2], Hmisc [9], and kableExtra [43]. We used R markdown [3] and bookdown [41] to generate web-enabled supplemental material. Our experimental data is available on the Open Science Framework at https://osf.io/xpjft/ [26].

Acknowledgements We thank members of the Michigan State University (MSU) Digital Evolution Laboratory for helpful comments and suggestions on this work. We thank the participants of the 2021 Genetic Programming in Theory and Practice workshop for lively discussion of our work. We especially thank Lee Spector for encouraging remarks and insightful feedback on our manuscript. MSU provided computational resources through the Institute for Cyber-Enabled Research. This work was supported in part by the National Science Foundation (NSF) through the BEACON Center (DBI-0939454) and a Graduate Research Fellowship to AL (DGE-1424871) and by the GEM Fellowship Program and Oak Ridge National Laboratory (ORNL). Any opinions, findings, and conclusions or recommendations expressed in this material are those of the author(s) and do not necessarily reflect the views of MSU, the NSF, GEM, or ORNL.

References

1. Aenugu, S., Spector, L.: Lexicase selection in learning classifier systems. In: Proceedings of the Genetic and Evolutionary Computation Conference on - GECCO '19, pp. 356–364. ACM Press, Prague, Czech Republic (2019)
2. Ahlmann-Eltze, C., Patil, I.: ggsignif: significance brackets for ggplot2. R package version 0.6.2. https://CRAN.R-project.org/package=ggsignif (2020)
3. Allaire, J., Xie, Y., McPherson, J., Luraschi, J., Ushey, K., Atkins, A., Wickham, H., Cheng, J., Chang, W., Iannone, R.: rmarkdown: dynamic documents for R. R package version 2.6. https://github.com/rstudio/rmarkdown (2020)
4. Dolson, E., Lalejini, A., Jorgensen, S., Ofria, C.: Interpreting the tape of life: ancestry-based analyses provide insights and intuition about evolutionary dynamics. Artif. Life 26, 58–79 (2020)
5. Dolson, E.L., Banzhaf, W., Ofria, C.: Ecological theory provides insights about evolutionary computation. preprint, PeerJ Preprints. https://peerj.com/preprints/27315 (2018). https://doi.org/10.7287/peerj.preprints.27315v1
6. Eiben, A.E., Schippers, C.A.: On evolutionary exploration and exploitation. Fundamenta Informaticae 35(1–4), 35–50 (1998)
7. Ferguson, A.J., Hernandez, J.G., Junghans, D., Lalejini, A., Dolson, E., Ofria, C.: Characterizing the effects of random subsampling on lexicase selection. In: Banzhaf, W., Goodman, E., Sheneman, L., Trujillo, L. (eds.) Genetic Programming Theory and Practice XVII, pp. 1–23. Springer (2020)
8. Garnier, S.: viridis: default color maps from matplotlib. R package version 0.5.1. https://github.com/sjmgarnier/viridis (2018)
9. Harrell Jr., F.E.: Hmisc: harrell miscellaneous. R package version 4.4-2. https://CRAN.R-project.org/package=Hmisc (2020)
10. Helmuth, T., Abdelhady, A.: Benchmarking parent selection for program synthesis by genetic programming. In: Proceedings of the 2020 Genetic and Evolutionary Computation Conference Companion, pp. 237–238 (2020)
11. Helmuth, T., Kelly, P.: PSB2: the second program synthesis benchmark suite. In: Proceedings of the Genetic and Evolutionary Computation Conference, pp. 785–794. ACM, Lille France (2021)
12. Helmuth, T., McPhee, N.F., Spector, L.: Effects of Lexicase and tournament selection on diversity recovery and maintenance. In: Proceedings of the 2016 on Genetic and Evolutionary Computation Conference Companion - GECCO '16 Companion, pp. 983–990. ACM Press, Denver, Colorado, USA (2016)

13. Helmuth, T., McPhee, N.F., Spector, L.: Lexicase selection for program synthesis: a diversity analysis. In: Riolo, R., Worzel, W., Kotanchek, M., Kordon, A. (eds.) Genetic Programming Theory and Practice XIII, pp. 151–167. Springer International Publishing, Cham (2016)
14. Helmuth, T., Pantridge, E., Spector, L.: On the importance of specialists for lexicase selection. Genetic Programming and Evolvable Machines (2020)
15. Helmuth, T., Spector, L.: General program synthesis benchmark suite. In: Proceedings of the 2015 on Genetic and Evolutionary Computation Conference - GECCO '15, pp. 1039–1046. ACM Press, Madrid, Spain (2015)
16. Helmuth, T., Spector, L.: Explaining and exploiting the advantages of down-sampled lexicase selection. In: The 2020 Conference on Artificial Life, pp. 341–349. MIT Press, Online (2020)
17. Helmuth, T., Spector, L.: Problem-solving benefits of down-sampled lexicase selection (2021). arXiv:2106.06085 [cs]
18. Helmuth, T., Spector, L., Matheson, J.: Solving uncompromising problems with lexicase selection. IEEE Trans. Evol. Comput. **19**(5), 630–643 (2015). https://doi.org/10.1109/TEVC.2014.2362729
19. Hernandez, J.G., Lalejini, A., Dolson, E., Ofria, C.: Random subsampling improves performance in lexicase selection. In: Proceedings of the Genetic and Evolutionary Computation Conference Companion, pp. 2028–2031 (2019)
20. Hernandez, J.G., Lalejini, A., Ofria, C.: Supplemental Material GitHub Repository (2021). https://doi.org/10.5281/zenodo.5020769
21. Hooker, J.N.: Testing heuristics: we have it all wrong. J. Heuristics **1**, 33–42 (1995)
22. Jundt, L., Helmuth, T.: Comparing and combining lexicase selection and novelty search. In: Proceedings of the Genetic and Evolutionary Computation Conference, pp. 1047–1055. ACM, Prague Czech Republic (2019)
23. Kassambara, A.: rstatix: pipe-friendly framework for basic statistical tests. R package version 0.7.0. https://rpkgs.datanovia.com/rstatix/ (2021)
24. La Cava, W., Helmuth, T., Spector, L., Moore, J.H.: A probabilistic and multi-objective analysis of lexicase selection and ε-lexicase selection. Evol. Comput. **27**, 377–402 (2019)
25. La Cava, W., Spector, L., Danai, K.: Epsilon-lexicase selection for regression. In: Proceedings of the Genetic and Evolutionary Computation Conference 2016, pp. 741–748 (2016)
26. Lalejini, A.M., Hernandez, J.G.: Experiment data. https://osf.io/xpjft/ (2021). https://doi.org/10.17605/OSF.IO/XPJFT
27. Lehman, J., Stanley, K.O.: Exploiting open-endedness to solve problems through the search for novelty. In: Proceedings of the Eleventh International Conference on Artificial Life (Alife XI). MIT Press (2008)
28. Lehman, J., Stanley, K.O.: Abandoning objectives: evolution through the search for novelty alone. Evol. Comput. **19**, 189–223 (2011)
29. Metevier, B., Saini, A.K., Spector, L.: Lexicase selection beyond genetic programming. In: Banzhaf, W., Spector, L., Sheneman, L. (eds.) Genetic Programming Theory and Practice XVI. Genetic and Evolutionary Computation, pp. 123–136. Springer International Publishing, Cham (2019). https://doi.org/10.1007/978-3-030-04735-1_7
30. Moore, J.M., McKinley, P.K.: A comparison of multiobjective algorithms in evolving quadrupedal gaits. In: Tuci, E., Giagkos, A., Wilson, M., Hallam, J. (eds.) From Animals to Animats 14, vol. 9825, pp. 157–169. Springer International Publishing, Cham (2016)
31. Moore, J.M., Stanton, A.: Lexicase selection outperforms previous strategies for incremental evolution of virtual creature controllers. In: Proceedings of the 14th European Conference on Artificial Life ECAL 2017, pp. 290–297. MIT Press, Lyon, France (2017)
32. Neuwirth, E.: RColorBrewer: colorbrewer palettes. R package version 1.1-2. https://CRAN.R-project.org/package=RColorBrewer (2014)
33. Ofria, C., Moreno, M.A., Dolson, E., Lalejini, A., Rodriguez-Papa, S., Fenton, J., Perry, K., Jorgensen, S., Hoffman, R., Miller, R., Edwards, O.B., Stredwick, J., G, N.C., Clemons, R., Vostinar, A., Moreno, R., Schossau, J., Zaman, L., Rainbow, D.: Empirical: a scientific software library for research, education, and public engagement (2020). https://doi.org/10.5281/zenodo.4141943

34. Orzechowski, P., La Cava, W., Moore, J.H.: Where are we now? A large benchmark study of recent symbolic regression methods. In: Proceedings of the Genetic and Evolutionary Computation Conference, pp. 1183–1190. ACM, Kyoto Japan (2018)
35. R Core Team: R: A language and environment for statistical computing. R Foundation for Statistical Computing, Vienna, Austria. https://www.R-project.org/ (2020)
36. Spector, L.: Assessment of problem modality by differential performance of lexicase selection in genetic programming: a preliminary report. In: Proceedings of the Fourteenth International Conference on Genetic and Evolutionary Computation Conference Companion - GECCO Companion '12, p. 401. ACM Press, Philadelphia, Pennsylvania, USA (2012)
37. Spector, L., Cava, W.L., Shanabrook, S., Helmuth, T., Pantridge, E.: Relaxations of lexicase parent selection. In: Banzhaf, W., Olson, R.S., Tozier, W., Riolo, R. (eds.) Genetic Programming Theory and Practice XV, pp. 105–120. Springer International Publishing, Cham (2018)
38. Wickham, H.: tidyverse: easily install and load the Tidyverse. R package version 1.3.0. https://CRAN.R-project.org/package=tidyverse (2019)
39. Wickham, H., Chang, W., Henry, L., Pedersen, T.L., Takahashi, K., Wilke, C., Woo, K., Yutani, H., Dunnington, D.: ggplot2: create elegant data visualisations using the grammar of graphics. R package version 3.3.4. https://CRAN.R-project.org/package=ggplot2 (2021)
40. Wilke, C.O.: cowplot: Streamlined plot theme and plot annotations for ggplot2. R package version 1.1.0. https://wilkelab.org/cowplot/ (2020)
41. Xie, Y.: bookdown: authoring books and technical documents with R markdown. R package version 0.21. https://github.com/rstudio/bookdown (2020)
42. Xie, Y.: knitr: A General-Purpose Package for Dynamic Report Generation in R. R package version 1.30. https://yihui.org/knitr/ (2020)
43. Zhu, H.: kableExtra: construct complex table with kable and pipe syntax. R package version 1.3.4. https://CRAN.R-project.org/package=kableExtra (2021)

Chapter 6
Feature Discovery with Deep Learning Algebra Networks

Michael F. Korns

Abstract Deep learning neural networks have produced some notable well publicized successes in several fields. Genetic Programming has also produced well publicized notable successes. Inspired by the deep learning successes with neural nets, we experiment with deep learning algebra networks where the network remains unchanged but where the neurons are replaced with general algebraic expressions. The training algorithms replace back propagation, counter propagation, etc. with a combination of genetic programming to generate the algebraic expressions and multiple regression, logit regression, and discriminant analysis to train the deep learning algebra network. These enhanced algebra networks are trained on ten theoretical classification problems with good performance advances which show a clear statistical performance improvement as network architecture is expanded.

6.1 Introduction

Deep learning neural networks have produced some notable successes in several fields [18–21, 23]. Inspired by the deep learning successes with neural nets, we extend our Abstract Regression Classification (ARC) system to evolve deep learning networks of algebraic expressions. These deep learning algebra networks are such that the network is unchanged but the neurons are replaced with general algebraic expressions. The new enhanced system is used to train algebra networks on ten theoretical classification problems with good performance advances. The performance advances are analyzed as the network architecture is expanded both by network depth (*i.e. number of hidden layers in the network*) and by the width of each network layer (*i.e. number of neurons in a layer*). Additionally, the algebra networks will be analyzed from the vantage point of feature discovery, with the layer width being viewed as multiple attempts at discovering the same feature, and the network depth being viewed as attempts to discover multiple new features.

M. F. Korns (✉)
Korns Associates, San Juan, PR 00911, USA
e-mail: mkorns@korns.com

W. Banzhaf et al. (eds.), *Genetic Programming Theory and Practice XVIII*,
Genetic and Evolutionary Computation,
https://doi.org/10.1007/978-981-16-8113-4_6

The problems we are attempting to solve herein are described by the simple matrix equation in Eq. 6.1 where Y is a numeric vector of N elements and X is a numeric matrix of N rows and M columns, $\mathbf{H_y}$ is an optimized function on X and error is the term to be minimized. A perfect score would be where $error = 0$.

$$Y = \mathbf{H_y}(X) + error \tag{6.1}$$

In this paper we will restrict our research to basic feed-forward deep learning neural networks with multiple hidden layers and a numeric output layer with a single neuron. For basic feed forward deep learning neural nets, each deep learning neural network is composed of an input layer (*with multiple inputs*), multiple hidden layers (*each with multiple self-similar neurons*), and a final output layer (*with one or more neurons*). The input layer is a collection of simple numeric values, while each of the hidden layers and the output layer are a collection of *simulated* neurons. Examining the first hidden layer of a simple feed forward neural network we discover a collection of activation function capped weighted sums of the inputs (*which form the simulated "neurons" in the hidden layer*). Each hidden layer of the network contains many of these simulated neurons [19]. If we arbitrarily choose the hyper tangent function, *for our activation function*, each of our first layer neurons can be expressed as a formula like the following.

$$H_{1j} = tanh(c_{1j0} + c_{1j1}X_1 + c_{1j2}X_2 + \cdots + c_{1jM}X_M) \tag{6.2}$$

The X_1 thru X_M represent the numeric input values from the input layer. The c_{1j0} thru c_{1jM} represent the weights. The term H_{1j} represents the value of the jth neuron in the first layer. There are J such weighted sums, simulated neurons, in the first hidden layer. As we can see, each hidden layer of the neural network contains many neurons and even more weights.

Examining the second layer of the neural network we encounter another collection of weighted sums, with inputs from the first layer neurons, which form the "neurons" in the second layer.

$$H_{2k} = tanh(c_{2k0} + c_{2k1}H_{11} + c_{2kj}H_{1j} + \cdots + c_{2kJ}H_{1J}) \tag{6.3}$$

The H_{11} thru H_{1J} represent the J neurons in the first layer which are inputs to the second layer neurons. The c_{2k0} thru $c|2KJ$ represent the weights. The term H_{2k} represents the value of the kth neuron in the second layer. There are K such weighted sums, simulated neurons, in the second hidden layer. This progression continues neuron by neuron, layer by layer until the final output neuron which is also a weighted sum, like the following formula.

$$Hy = tanh(c_{O0} + c_{O1}H_{N1} + c_{Oq}H_{Nq} + \cdots + c_{OQ}H_{NQ}) \tag{6.4}$$

The H_{N1} thru H_{NQ} represent the Q neurons in the output layer which are inputs to the final hidden layer neurons. The c_{O0} thru c_{OQ} represent the weights. The term H_{Nq}

represents the value of the qth neuron in the output layer. There are Q such weighted sums, simulated neurons, in the output layer and the output function, **Hy**, produces one numeric output value from the output layer. Deep learning neural networks often have a large number of neurons in each layer and many layers—often tens and even hundreds of layers or more. Obviously, as the number of neurons and layers grows, we can easily have an explosively large number of formulas such as Eqs. 6.2–6.4 with thousands of weights.

In this paper we view the network from a *feature discovery* vantage point, let us describe the set of all input values and all neurons in the network as follows.

$$S = \{X_1, X_2, \ldots, X_M, H_{11}, \ldots, H_{1J}, H_{21}, \ldots, H_{1N}, \ldots, H_{O1}, \ldots, H_{OQ}, H_y, Y\} \tag{6.5}$$

Expressed as in Eq. 6.5, S is the set of all input features and all hidden layer neurons including the output layer neuron H_y, and the target variable Y. S may be quite a large set, and it is comprehensive. From S we can derive all of the numeric values of the inputs and each hidden layer neuron, plus the values from output neuron. In a batch supervised learning context, the numeric values (from the inputs and the neurons) form an array of rows and columns with each row being a training point and each column being the values of the inputs, simulated neurons, and the output. During training, all neurons in the network have their coefficient weights altered to optimize Eq. 6.1.

From S we can infer the dependency properties of the network (*even though S does not contain information about the physical layout of the network*). For instance, a simple dependency graph of the inputs to every neuron formula will identify the layers in S (any two neurons are in the same layer if they have identical dependency sets). From the dependency graph we can tell if the network is acyclic (such as a feed forward network) or if the network contains cycles (i.e. a feedback or other more complex network). Whether there is a single output or multiple outputs can also be determined from the collection of neuron formulas. Once we have the set, S, we no longer need the physical network to train and/or compute the network output—especially if we are focused on analyzing the network from a feature discovery vantage point. In this paper we will be adapting all feed forward, acyclic, single output, neural networks as similarly structured deep learning algebra networks.

Furthermore, examining the set S we can view the neural net hidden layers as a form of new feature discovery. Each hidden layer neuron is another new feature, added to the list of original input features, and available as inputs to selected other up layer neurons based upon the physical network topology. The multiple neuron formulas in any given layer are analogous to repeated attempts to find relevant new features from the inputs available to the specified layer (remember all neurons in a layer have identical dependency sets). Therefore, they represent repeated attempts to discover new features from the same input data elements but with different randomized learning parameters.

Examining the simulated neurons in Eqs. 6.2–6.4 we see that they are quite specialized and restricted. A great deal of research has gone into enhancing the basic neuron formula to make it less restrictive, while still keeping the claim to biological inspiration [19]. It is inevitable that we may wonder, "What would result from making these simple simulated neuron formulas more general?" For instance, we might want to create a recurrent neural network where the neurons have temporal state [19]. There are many reasons to generalize the simulated neurons in a deep learning neural network. Assuming one is willing to relinquish the biologically inspired claim, the most obvious way to generalize these simulated neuron formulas is to make them activation function capped algebraic general linear models [15] like the following formula.

$$H_{n+1p} = F_{n0}\left(c_{n0} + c_{n1}F_{n1}(S) + c_{np}F_{np}(S) + \cdots + c_{nP}F_{nP}(S)\right) \tag{6.6}$$

This formula expresses the generalized linear formula for the pth generalized neuron in the $n + 1st$ hidden layer. If F_{n0} represents the hyper tangent function and each F_{np} represents the pth projection function, then H_{n+1p} is our original simple simulated neuron. However, with proper function substitutions, H_{n+1p} can be any algebraic formula we wish—an algebraic neuron. For instance, the following algebraic neuron is just one of a countably infinite number of formulas which can be used to represent our new algebraic neurons.

$$H_{n+1p} = sin(c_{n0} + c_{n1}cos\left(\frac{H_{21}}{X_1}\right) + c_{np}^2[\text{if}(H_{81} < H_{48}, H_{29}, X_{22})) + \cdots + c_{nP}log(H_{67})] \tag{6.7}$$

As one can easily see, the two equations are very similar to Eq. (6.4), the simulated neuron, being a specific restricted case of the more general algebraic neuron of Eq. 6.6. Algebra neurons form a general class of neural expressions of which the more restricted simulated neuron is a subset. We can also retain or drop the activation function with algebra neurons without loss of generality as follows.

$$H_{n+1p} = c_{n0} + c_{n1}c_{n1}F_{n1}(S) + c_{np}F_{np}(S) + \cdots + c_{nP}F_{nP}(S) \tag{6.8}$$

While the restricted neural net expressions (6.4) are biologically inspired, they are almost always quite verbose. This verbose property is largely responsible for the neural net's reputation as a black box learning methodology. For instance, if the actual correct answer to a hypothetical regression problem is a simple sine wave.

$$y = c_0 + sin(X_{21}) \tag{6.9}$$

It will take around ten or more nested basic neuron expressions of format (6.4) to simulate this simple sine wave *and* it will be unclear to most human readers what the multiple nested restricted neurons are trying to accomplish. Whereas the more general algebraic neuron solution is literally the expression of Eq. 6.9. It is this terse quality which allows networks of algebraic neurons to be a more human readable white box learning methodology. On average, it can often take approximately one hundred to one thousand simple neurons to simulate a single modestly complex algebraic neuron.

Each Abstract Regression Classification (ARC) network algebraic neuron has one output signal which may be input multiple times to the algebraic neurons in the layers above. Such algebra neurons can drop the activation function as the remaining generalized linear model will be no less general [15], or they can keep the activation function. A simple ARC network might appear as follows.

$$\text{inputs: } x_1, x_2, \ldots, x_M \tag{6.10}$$

$$\text{hidden neuron: } h_1 = c_{10} + c_{11} * (x_1) + c_{12} * (cos(x_{21})) + \cdots + c_{1M} * \left(\frac{x_3}{x_6}\right) \tag{6.11}$$

Neural networks can learn in an unsupervised or a supervised setting. In this paper we are concerned only with batch supervised learning. Neural network supervised training is performed by a selection of learning algorithms which can be applied "batch" or "online" (i.e. back-propagation, counter-propagation, or RProp to name few [19]). Most of the popular neural net training algorithms incrementally modify the entire collection of weights in, $\mathbf{H_y}$, trickling down incrementally, layer by layer so as to optimize the final output $\mathbf{H_y}$.

ARC Network training proceeds bottom up, layer by layer, where each neuron, Hoq, is successively optimized against Y. Training ARC networks is predicated on the fact that algebraic neurons are delivered in the format of general linear models (GLM) [15]. General linear models are amenable to four types of machine learning which we use extensively in ARC. These are (a) genetic programming [13] for evolving concrete algebraic neuron formulas to be optimized, (b) multiple regression [22] for optimizing algebraic neuron formulas with numeric outputs, (c) logistic regression [16] for optimizing algebraic neuron formulas with binary outputs, and (c) linear discriminant analysis [3, 14] for optimizing algebraic neuron formulas with m-class outputs. The ARC learning algorithm employs industrialized versions of these four learning algorithms, as described in [11], and proceeds bottom up, one algebraic neuron expression at a time, adjusting that neuron expression to optimize equation (E0), then proceeding up through the dependency network hierarchy until the final output expression is optimized, Hy. Interestingly, one could define an entire ARC network of restricted simulated neuron algebra expressions and train it in this bottom-up approach. We will not explore this avenue here; but it would offer an experimental mechanism for comparing the four bottom-up ARC training algorithms versus the several popular trickle-down neural network training algorithms.

Neural network architecture determines the total number of neurons in a neural network. Neurons such as Eqs. 6.2–6.4 are fixed unchanging concrete formulas wherein only the coefficient weights change during training. One can examine a neural network architecture in advance and compute the maximum number of neurons that will ever be optimized. Conversely, the action of genetic programming makes each ARC network an abstract network rather than a fixed concrete network. Algebra neurons like Eqs. 6.6 and 6.8, together with genetic programming technology, can best be thought of as neuron factories which act upon abstract neurons such as Eq. 6.6 and produce multiple concrete neurons such as Eq. 6.7. ARC networks are abstract in nature—composed of neuron factories rather than concrete neurons. During ARC network training, each abstract neuron factory produces hundreds and thousands of concrete neurons whose coefficients are then optimized against Y. So not only is each algebra neuron, on average, far more complex than each neural network neuron, but many hundreds and thousands of concrete algebra neurons are produced and optimized during ARC network training for every single abstract neuron in the ARC network architecture. Even a small ARC network can contain tens of thousands of concrete optimized algebra neurons. Larger ARC networks can contain millions of concrete optimized algebra neurons.

Eliminating poor performing neurons (pruning) has been shown to be a vital technique for enhancing neural net training [1]. We have found that pruning of algebra neurons also enhances ARC network training. In fact, given the large number of algebra neurons produced in training each ARC network, pruning of poor performing algebra neurons is essential to increase learning efficiency and to reduce bloat. ARC algebraic neurons can be pruned based upon any or more of the following: coefficient strength, principal component analysis, and fitness competition. In this paper, all three pruning methodologies are employed to the highest levels possible.

Of course, the general nature of the algebraic neuron places the claim of biological inspiration in jeopardy. Arguments have been made that certain enhancements of the simple neuron are biologically inspired. For instance, some scientists have argued that recurrent neurons are biologically inspired. However, few scientists would argue that the more general algebraic neuron is biologically inspired. In the general algebraic case, the term neuron is more of a legacy nomenclature than a claim to neuron-similar behavior. So, if algebraic neurons are not biological inspired, can networks of algebraic neurons be trained and can they be made useful in any practical sense?

This paper includes a performance comparison of deep learning algebra networks and five well-known commercially available classification algorithms on ten theoretical noiseless classification problems. The five commercially available classification algorithms are found in the KNIME system [14], and are as follows: Multiple Layer Perceptron Learner (MLP); Decision Tree Learner (DTL); Random Forest Learner (RFL); Tree Ensemble Learner (TEL); and Gradient Boosted Trees Learner (GBTL). We show that, on the theoretical problems, the two best classification algorithms are Gradient Boosted Decision Trees (GBTL) and this paper's deep learning algebra networks (ARC). Furthermore, we show that the performance across all ten theoretical problems consistently improves as the algebra network is expanded both in width (neurons per layer) and depth (number of layers).

6.2 ARC Background

By way of providing some background, our Abstract Regression Classification (ARC) system has been under research and development since 2004 [5–12]. ARC has been heavily industrialized and requires no genetic programming specific input parameters. Only the names of the training and testing data files and the nature of the target variable (numeric, binary, or nary) need be specified. The selection of the fitness method, running of multiple genetic programming runs with different random number seeds, splicing the different runs together to form a layered network, determining when the system is finished learning, etc., all of these tasks are hidden from the user by the ARC system planning module. Only a single ARC training run is needed per problem, and the ARC system is guaranteed to converge on the best solution it can find in the finite time and computation resources allotted.

The ARC planning module is based around the Regression Query Language (RQL) which is an SQL inspired search language for specifying genetic programming symbolic regression and classifications runs. The RQL language is briefly described in [7] and can be used to set in motion single island or multiple island genetic programming runs with aged-layered, pareto, elitist, and many other GP methodologies. The RQL language employs industrialized implementations of (a) genetic programming [13, 21] for evolving algebraic neuron formulas to be optimized, (b) multiple regression [22] for optimizing algebraic neuron formulas with numeric outputs, (c) logistic regression [16] for optimizing algebraic neuron formulas with binary outputs, and (c) linear discriminant analysis [3, 14] for optimizing algebraic neuron formulas with m-class outputs. The RQL language is quite sophisticated. For instance, one study includes an RQL specification which is conjectured to be absolutely accurate on certain scientific problems [7]. The ARC planning module currently contains a library of numerous predefined RQL searches known to be effective for specific problems. The planning module applies its library of known RQL searches, based upon its own heuristic and statistical analysis of the data to be optimized. Each ARC training run hides thousands of separate genetic programming runs from the user. Human intervention is not required.

6.3 Regression in Brief

Regression, single and multiple, involves a single dependent variable and one or more independent variables. It is a statistical technique that develops an optimal mathematical relationship between one or more real independent variables and a real dependent variable. Most modern regression tools manage linear regression. Symbolic regression tools attack the mathematical problem of nonlinear regression by employing genetic programming.

The canonical generalization of linear regression into nonlinear regression is the class of Generalized Linear Models (GLMs) as described in [15]. A GLM is a linear combination of I algebraic functions $B_i; i = 0, \ldots, I$, a dependent variable y, and an independent data point with M features $x = (x_0, x_1, x_2, \ldots, x_{M-1})$: such that

$$y = \gamma(x) = c_0 + c_1 B_1(x) + c_2 B_2(x) + \cdots + c_{M-1} B_{M-1}(x) + \text{error} \quad (6.12)$$

As a broad generalization, GLMs can represent any possible nonlinear formula. However, the format of the GLM makes it amenable to existing linear regression theory and tools since the GLM model is linear on each of the algebraic functions B_i (although each algebraic function may be nonlinear). For a given vector of dependent variables, Y, and a vector of independent data points, X, symbolic regression will search for a set of algebraic functions and coefficients which minimize error. In [13] the algebraic functions selected by symbolic regression will be formulas as in the following examples:

(E12) B0 = x3 (E13) B1 = x1+x4 (E14) B2 = sqrt(x2)/tan(x5/4.56) (E15) B2 = sqrt(x2)/tan(x5/4.56)

$$B_0 = x_3 \tag{6.13}$$

$$B_1 = x_1 + x_4 \tag{6.14}$$

$$B_2 = \frac{\sqrt{x_2}}{tan(\frac{x_5}{4.56})} \tag{6.15}$$

$$B_3 = tanh[cos(x_2 * 0.2) * (x_5 + abs(x_1))^3] \tag{6.16}$$

Once a suitable set of algebraic functions B have been selected (via genetic programming), we can discover the proper set of coefficients C deterministically using standard simple or multiple regression [22]. The value of the GLM model is that one can use standard regression techniques and theory to optimize for the constants while using genetic programming to search for the optimal algebraic functions.

6.4 Classification in Brief

For all binary classification, we use Logit Regression (LOG) as in [16]. For all M-Class classification, we use Linear Discriminant Analysis (LDA). Linear discriminant analysis is a generalization of Fischer's linear discriminant, which is a method to find a linear combination of features which best separates K classes of training points [3, 11, 14]. Both LOG and LDA are used extensively in Statistics, Machine Learning, and Pattern Recognition.

In symbolic classification problems, an N by M matrix of independent training points, X, is matched with an N vector of dependent values containing only categorical unordered values between 1 and K. The fitness measure is the classification error percent (CEP). Linear discriminant analysis is employed as the assisted fitness training technique in our ARC system. The CEP is the percent of misclassified testing points (i.e. the count of misclassifications divided by the number of testing points).

Our symbolic classification system outputs K predictor functions (one for each class). These functions are called discriminants, $D_k(X) \approx Y_k$, and there is one discriminant function for each class. The format of ARC's discriminant function output is always as follows.

$$y = argmax(D_1, D_2, \ldots, D_K) \quad (6.17)$$

The argmax function returns the class index for the largest valued discriminant function. For instance if $D_i = max(D_1, D_2, \ldots, D_K)$, then $i = argmax(D_1, D_2, \ldots, D_K)$.

A central aspect of LDA is that each discriminant function is a linear variation of every other discriminant function. For instance, if the ARC symbolic classification system produces a candidate with B algebraic neuron functions, then each discriminant function has the following format:

$$\begin{aligned} D_0 &= c_{00} + c_{01} * Bf_1 + c_{02} * Bf_2 + \cdots + c_{0B} * Bf_B \\ D_1 &= c_{10} + c_{11} * Bf_1 + c_{12} * Bf_2 + \cdots + c_{1B} * Bf_B \\ D_k &= c_0 + c_{k1} * Bf_1 + c_{k2} * Bf_2 + \cdots + c_{kB} * Bf_B \end{aligned}$$

The $K * (B + 1)$ coefficients are selected so that the ith discriminant function has the highest value when the $y = i$ (i.e. the class is i). The industrialized LDA technology ARC uses for selecting these optimized coefficients c_{00} to c_{KB} is discussed in more detail here [10].

6.5 Industrial Regression Classification

The single, multiple, and logit regression plus the linear discriminant analysis algorithms in ARC have been industrialized to handle real world problems. These quite exacting deterministic algorithms all suffer from their requirement that certain assumptions about the training data hold true—namely that the data have a normal distribution, that the training matrices be nonsingular, etc. In cases where the data does not strictly conform to these assumptions, these deterministic algorithms can fail or produce inaccurate results. Whenever ARC discovers poorly structured training data, the deterministic regression classification algorithms are forced into approximately accurate coefficients. Next a layer of fast evolutionary algorithms is applied to coerce the approximately accurate coefficients into a more accurate set of coefficients. These evolutionary algorithms include modified sequential minimal optimization and bees swarm optimization [4, 17]. These industrial enhancements create regression classification algorithms which are robust even with poorly formed training data.

6.6 Theoretical Test Problems—Classification

A set of ten artificial classification problems are constructed, with no noise, to test the efficacy of the new ARC deep learning networks. Each of the artificial test problems is created around an X training matrix filled with random real numbers in the range $[-10.0, +10.0]$. The number of rows and columns in each test problem varies from 5000×25 to 5000×1000 depending upon the difficulty of the problem. The number of classes varies from $Y = 0, 1$ to $Y = 0, 1, 2, 3, 4$ depending upon the difficulty of the problem. The test problems are designed to vary from extremely easy to very difficult. The first test problem is linearly separable with 2 classes on 25 columns. The tenth test problem is nonlinear multimodal with 5 classes on 1000 columns.

Standard statistical best practices out of sample testing are employed. First the training matric X is filled with random real numbers in the range $[-10.0, +10.0]$, and the Y class values are computed from the argmax functions specified below. A champion is trained on the training data. Next a testing matrix X is filled with random real numbers in the range $[-10.0, +10.0]$, and the Y class values are computed from the argmax functions specified below.

The argmax functions used to create each of the ten artificial test problems are as follows:

- C_1: $y = argmax(D_1, D_2)$ where $Y = 1, 2$, X is 5000×25, and each D_i is as follows:

$$\begin{cases} D_1 = 1.57x_0 - 39.34x_1 + 2.13x_2 + 46.59x_3 + 11.54x_4 \\ D_2 = -1.57x_0 + 39.34x_1 - 2.13x_2 - 46.59x_3 - 11.54x_4 \end{cases}$$

- C_2: $y = argmax(D_1, D_2)$ where $Y = 1, 2$, X is 5000×100, and each D_i is as follows:

$$\begin{cases} D_1 = 5.16x_0 - 19.83x_1 + 19.83x_2 + 29.31x_3 + 5.29x_4 \\ D_2 = -5.16x_0 + 19.83x_1 - 0.93x_2 - 29.31x_3 + 5.29x_4 \end{cases}$$

- C_3: $y = argmax(D_1, D_2)$ where $Y = 1, 2$, X is $5000 \times 1{,}000$, and each D_i is as follows:

$$\begin{cases} D_1 = -34.16x_0 + 2.19x_1 - 12.73x_2 + 5.62x_3 - 16.36x_4 \\ D_2 = 34.16x_0 - 2.19x_1 + 12.73x_2 - 5.62x_3 + 16.36x_4 \end{cases}$$

- C_4: $y = argmax(D_1, D_2, D_3)$ where $Y = 1, 2, 3$, X is $5{,}000 \times 25$, and each D_i is as follows:

$$\begin{cases} D_1 = 1.57\cos x_0 - 39.34x_{10}^2 + 2.13\dfrac{x_2}{x_3} + 46.59x_{13}^3 - 11.54\log x_4 \\ D_2 = -0.56\cos x_0 + 9.34x_{10}^2 + 5.28\dfrac{x_2}{x_3} - 6.10x_{13}^3 + 1.48\log x_4 \\ D_3 = 1.37\cos x_0 + 3.62x_{10}^2 + 4.04\dfrac{x_2}{x_3} + 1.95x_{13}^3 + 9.54\log x_4 \end{cases}$$

- C_5: $y = argmax(D_1, D_2, D_3)$ where $Y = 1, 2, 3$, X is 5,000 ×100, and each D_i is as follows:

$$\begin{cases} D_1 = 1.57\sin x_0 - 39.34x_{10}^2 + 2.13\dfrac{x_2}{x_3} + 46.59x_{13}^3 - 11.54\log x_4 \\ D_2 = -0.56\sin x_0 + 9.34x_{10}^2 + 5.28\dfrac{x_2}{x_3} - 6.10x_{13}^3 + 1.48\log x_4 \\ D_3 = 1.37\sin x_0 - 3.62x_{10}^2 + 4.04\dfrac{x_2}{x_3} + 1.95x_{13}^3 - 9.54\log x_4 \end{cases}$$

- C_6: $y = argmax(D_1, D_2, D_3)$ where $Y = 1, 2, 3$, X is 5,000 × 1,000, and each D_i is as follows:

$$\begin{cases} D_1 = 1.57\tanh x_0 - 39.34\sqrt{x_{10}} + 2.1\dfrac{x_2}{x_3} + 46.59x_{13}^3 - 11.54\log x_4 \\ D_2 = -0.56\tanh x_0 + 9.34\sqrt{x_{10}} + 5.28\dfrac{x_2}{x_3} - 6.10x_{13}^3 + 1.48\log x_4 \\ D_3 = 1.37\tanh x_0 - 3.62\sqrt{x_{10}} + 4.04\dfrac{x_2}{x_3} + 1.95x_{13}^3 - 9.54\log x_4 \end{cases}$$

- C_7: $y = argmax(D_1, D_2, D_3, D_4, D_5)$ where $Y = 1, 2, 3, 4, 5$, X is 5,000 × 25, and each D_i is as follows:

$$\begin{cases} D_1 = 1.57\cos\left(\dfrac{x_0}{x_{21}}\right) + 9.34x_{10}^2\dfrac{x_6}{x_{14}} + 2.13\dfrac{x_2}{x_3}\log x_8 + 46.59x_{13}^3\dfrac{x_9}{x_2} - 11.54\log(x_4x_{10}x_{15}) \\ D_2 = -1.56\cos\left(\dfrac{x_0}{x_{21}}\right) + 7.34x_{10}^2\dfrac{x_6}{x_{14}} + 5.28\dfrac{x_2}{x_3}\log x_8 + 6.10x_{13}^3\dfrac{x_9}{x_2} + 1.48\log(x_4x_{10}x_{15}) \\ D_3 = 2.31\cos\left(\dfrac{x_0}{x_{21}}\right) + 12.34x_{10}^2\dfrac{x_6}{x_{14}} - 1.28\dfrac{x_2}{x_3}\log x_8 + 0.21x_{13}^3\dfrac{x_9}{x_2} + 2.61\log(x_4x_{10}x_{15}) \\ D_4 = -0.56\cos\left(\dfrac{x_0}{x_{21}}\right) + 8.34x_{10}^2\dfrac{x_6}{x_{14}} + 16.71\dfrac{x_2}{x_3}\log x_8 - 2.93x_{13}^3\dfrac{x_9}{x_2} + 5.228\log(x_4x_{10}x_{15}) \\ D_5 = 1.07\cos\left(\dfrac{x_0}{x_{21}}\right) - 1.62x_{10}^2\dfrac{x_6}{x_{14}} - 0.04\dfrac{x_2}{x_3}\log x_8 - 0.95x_{13}^3\dfrac{x_9}{x_2} + 0.54\log(x_4x_{10}x_{15}) \end{cases}$$

- C_8: $y = argmax(D_1, D_2, D_3, D_4, D_5)$ where $Y = 1, 2, 3, 4, 5$, X is 5,000 × 100, and each D_i is as follows:

$$\begin{cases} D_1 = 1.57 \sin\left(\frac{x_0}{x_{11}}\right) + 9.34 x_{12}^2 \frac{x_{46}}{x_4} + 2.13 \frac{x_{21}}{x_3} \log x_{18} + 46.59 x_3^3 \frac{x_9}{x_2} - 11.54 \log(x_{14} x_{10} x_{15}) \\ D_2 = -1.56 \sin\left(\frac{x_0}{x_{11}}\right) + 7.34 x_{12}^2 \frac{x_{46}}{x_4} + 5.28 \frac{x_{21}}{x_3} \log x_{18} + 6.10 x_3^3 \frac{x_9}{x_2} + 1.48 \log(x_{14} x_{10} x_{15}) \\ D_3 = 2.31 \sin\left(\frac{x_0}{x_{11}}\right) + 12.34 x_{12}^2 \frac{x_{46}}{x_4} - 1.28 \frac{x_{21}}{x_3} \log x_{18} + 0.21 x_3^3 \frac{x_9}{x_2} + 2.61 \log(x_{14} x_{10} x_{15}) \\ D_4 = -0.56 \sin\left(\frac{x_0}{x_{11}}\right) + 8.34 x_{12}^2 \frac{x_{46}}{x_4} + 16.71 \frac{x_{21}}{x_3} \log x_{18} - 2.93 x_3^3 \frac{x_9}{x_2} + 5.228 \log(x_{14} x_{10} x_{15}) \\ D_5 = 1.07 \sin\left(\frac{x_0}{x_{11}}\right) - 1.62 x_{12}^2 \frac{x_{46}}{x_4} - 0.04 \frac{x_{21}}{x_3} \log x_{18} - 0.95 x_3^3 \frac{x_9}{x_2} + 0.54 \log(x_{14} x_{10} x_{15}) \end{cases}$$

- C_9: $y = argmax(D_1, D_2, D_3, D_4, D_5)$ where $Y = 1, 2, 3, 4, 5$, X is 5,000 × 1,000, and each D_i is as follows:

$$\begin{cases} D_1 = 1.57 \sin(x_{20} x_{11}) + 9.34 \tanh(\frac{x_{12}}{x_4} x_{46}) + 2.13 (x_{321} - x_3) \tan x_{18} + 46.59 \frac{x_3^2}{x_{49} x_{672}} \\ - 11.54 \log(x_{24} x_{120} x_{925}) \\ D_2 = -1.56 \sin(x_{20} x_{11}) + 7.34 \tanh(\frac{x_{12}}{x_4} x_{46}) + 5.28 (x_{321} - x_3) \tan x_{18} + 6.10 \frac{x_3^2}{x_{49} x_{672}} \\ + 1.48 \log(x_{24} x_{120} x_{925}) \\ D_3 = 2.31 \sin(x_{20} x_{11}) + 12.34 \tanh(\frac{x_{12}}{x_4} x_{46}) - 1.28 (x_{321} - x_3) \tan x_{18} + 0.21 \frac{x_3^2}{x_{49} x_{672}} \\ + 2.61 \log(x_{24} x_{120} x_{925}) \\ D_4 = -0.56 \sin(x_{20} x_{11}) + 8.34 \tanh(\frac{x_{12}}{x_4} x_{46}) + 16.71 (x_{321} - x_3) \tan x_{18} - 2.93 \frac{x_3^2}{x_{49} x_{672}} \\ + 5.228 \log(x_{24} x_{120} x_{925}) \\ D_5 = 1.07 \sin(x_{20} x_{11}) - 1.62 \tanh(\frac{x_{12}}{x_4} x_{46}) - 0.04 (x_{321} - x_3) \tan x_{18} - 0.95 \frac{x_3^2}{x_{49} x_{672}} \\ + 0.54 \log(x_{24} x_{120} x_{925}) \end{cases}$$

- C_{10}: $y = argmax(D_1, D_2, D_3, D_4, D_5)$ where $Y = 1, 2, 3, 4, 5$, X is 5,000 × 1,000, and each D_i is as follows:

$$\begin{cases} D_1 = 1.57A + 9.34 \tanh(\frac{x_{12}}{x_4} x_{46}) + 2.13B + 46.59 \frac{x_3^2}{x_{49} x_{672}} - 11.54 \log(x_{24} x_{120} x_{925}) \\ D_2 = -1.56A + 7.34 \tanh(\frac{x_{12}}{x_4} x_{46}) + 5.28B + 6.10 \frac{x_3^2}{x_{49} x_{672}} + 1.48 \log(x_{24} x_{120} x_{925}) \\ D_3 = 2.31A + 12.34 \tanh(\frac{x_{12}}{x_4} x_{46}) - 1.28B + 0.21 \frac{x_3^2}{x_{49} x_{672}} + 2.61 \log(x_{24} x_{120} x_{925}) \\ D_4 = -0.56A + 8.34 \tanh(\frac{x_{12}}{x_4} x_{46}) + 16.71B - 2.93 \frac{x_3^2}{x_{49} x_{672}} + 5.228 \log(x_{24} x_{120} x_{925}) \\ D_5 = 1.07A - 1.62 \tanh(\frac{x_{12}}{x_4} x_{46}) - 0.04B - 0.95 \frac{x_3^2}{x_{49} x_{672}} + 0.54 \log(x_{24} x_{120} x_{925}) \end{cases}$$

with

$$A = \begin{cases} \sin(x_{20}x_{11}), & \text{if } x_0 < x_{23} \\ \cos(x_{10}, & \text{otherwise} \end{cases}$$

and

$$B = \begin{cases} (x_{321} - x_3)\tan x_{18}, & \text{if } x_{10} < 0 \\ (x_{156} - x_{31})/\tanh x_{21}, & \text{otherwise} \end{cases}$$

6.7 Base Performance on the Theoretical Classification Problems

Here we compare the out of sample CEP testing scores of five well-known commercially available classification algorithms to determine where basic 1-layer ARC networks rank in competitive comparison. The five commercially available classification algorithms are available in the KNIME system [2], and are as follows: Multiple Layer Perceptron Learner (MLP); Decision Tree Learner (DTL); Random Forest Learner (RFL); Tree Ensemble Learner (TEL); and Gradient Boosted Trees Learner (GBTL). The following table lists each classification algorithm in descending order of average CEP scores on all ten theoretical test problems. The lower the CEP the more accurate the classification results. The ARCN1 network is composed of 1 hidden layer which has 5 algebra neuron factories (width = 5, depth = 1 – which is to say almost no network and just one training run with five neuron factories) (Fig. 6.1).

Test	MLP	DTL	TEL	RFL	ARC^{N1}	GBTL
C1	0.0072	0.0724	0.0496	0.0492	0.0138	0.0308
C2	0.0360	0.0740	0.0648	0.0664	0.0116	0.0240
C3	0.0724	0.0972	0.1526	0.1522	0.0132	0.0332
C4	0.0472	0.0174	0.0252	0.0260	0.0194	0.0170
C5	0.3250	0.0858	0.0946	0.0920	0.0712	0.0530
C6	0.6166	0.5396	0.6284	0.6286	0.5420	0.3198
C7	0.4598	0.2834	0.2284	0.2292	0.2272	0.2356
C8	0.4262	0.2956	0.2248	0.2250	0.2302	0.2340
C9	0.6904	0.6058	0.4334	0.4344	0.4188	0.4286
C10	0.5966	0.5966	0.4352	0.4296	0.4186	0.4286
Avg	0.3277	0.2667	0.2337	0.2332	0.2169	0.1804

Fig. 6.1 Test problem CEP testing results before deep learning enhancements

The top performer overall by a very small margin is the Gradient Boosted Trees Learner (GBTL). The penultimate performer is the ARCN1 1-Layer algebra network. The base ARCN1 network is composed of 1 layer with 5 algebra neuron factories producing an average of 51.6 K concrete algebra neurons per test case. Interestingly, the base ARC network performs reasonably well before deep learning enhancements.

6.8 Thin 2-Layer ARC Performance on the Theoretical Classification Problems

Here we compare the performance of a thin 2-Layer ARC network with the out of sample CEP testing scores of five well-known commercially available classification algorithms to determine where a 2-layer ARC network ranks in competitive comparison. The following table lists each classification algorithm in descending order of average CEP scores on all ten theoretical test problems. The lower the CEP the more accurate the classification results. The ARCN2 network is composed of 2 hidden layers each of which has 40 algebra neuron factories for a total of 80 algebra neuron factories in the entire network (Fig. 6.2).

The top performer overall by a reasonable margin is now the ARCN2 thin 2-Layer algebra network (width = 40, depth = 2). The Gradient Boosted Trees Learner (GBTL) has fallen behind. Interestingly, adding just two hidden layers and a total of 80 algebra neuron factories (40 algebra neuron factories per layer producing an average of 188.9 K concrete algebra neurons per test case) was enough to boost performance beyond that of the Gradient Boosted Trees Learner (GBTL).

Test	MLP	DTL	TEL	RFL	ARC^{N1}	GBTL	ARC^{N2}
C1	0.0072	0.0724	0.0496	0.0492	0.0138	0.0308	0.0004
C2	0.0360	0.0740	0.0648	0.0664	0.0116	0.0240	0.0004
C3	0.0724	0.0972	0.1526	0.1522	0.0132	0.0332	0.0022
C4	0.0472	0.0174	0.0252	0.0260	0.0194	0.0170	0.0158
C5	0.3250	0.0858	0.0946	0.0920	0.0712	0.0530	0.0490
C6	0.6166	0.5396	0.6284	0.6286	0.5420	0.3198	0.2518
C7	0.4598	0.2834	0.2284	0.2292	0.2272	0.2356	0.2264
C8	0.4262	0.2956	0.2248	0.2250	0.2302	0.2340	0.2238
C9	0.6904	0.6058	0.4334	0.4344	0.4188	0.4286	0.4142
C10	0.5966	0.5966	0.4352	0.4296	0.4186	0.4286	0.4160
Avg	0.3277	0.2667	0.2337	0.2332	0.2169	0.1804	0.1600

Fig. 6.2 Test problem CEP testing results after thin 2-layer deep learning enhancements

6.9 Ultrathin 8-Layer ARC Performance on the Theoretical Classification Problems

Here we compare the performance of an ultrathin 8-Layer ARC network with the out of sample CEP testing scores of five well-known commercially available classification algorithms to determine where an 8-layer ARC network ranks in competitive comparison. The following table lists each classification algorithm in descending order of average CEP scores on all ten theoretical test problems. The lower the CEP the more accurate the classification results. The ARCN3 network is composed of 8 hidden layers each of which has 10 algebra neuron factories for a total of 80 algebra neuron factories in the entire network (width = 10, depth = 8). On average the ARCN3 network's 80 neuron factories produced 1.4 M concrete neurons per test case (Fig. 6.3).

The ARCN2 network and the ARCN3 network have the same total number of algebra neuron factories. A comparison of their results highlights the difference between more thin layers or fewer wide layers. A win for the ARCN3 network would indicate that new feature discovery is important for network performance, while a win for the ARCN2 network would indicate that mere repetition is important for network performance. Notice that the dependency set for each of the layer 1 algebra neurons is the set X. While the dependency set for all of the layer 2 algebra neurons is the set $X \cup H_1$ where H_1 is the output of all of the layer 1 algebra neurons after pruning. The difference between the ARCN2 network and the ARCN3 network is that the layer 2 thru 8 algebra neurons in the ARCN3 network have access to the outputs of more previous neurons because the network is deep instead of wide.

The top performer overall by a slight margin is now the ARCN3 ultrathin 8-Layer algebra network (even though it has exactly the same total number of algebra neurons as the ARCN2 network). Whether this slight advantage persists for other

Test	MLP	DTL	TEL	RFL	ARC^{N1}	GBTL	ARC^{N2}	ARC^{N3}
C1	0.0072	0.0724	0.0496	0.0492	0.0138	0.0308	0.0004	0.0000
C2	0.0360	0.0740	0.0648	0.0664	0.0116	0.0240	0.0004	0.0016
C3	0.0724	0.0972	0.1526	0.1522	0.0132	0.0332	0.0022	0.0026
C4	0.0472	0.0174	0.0252	0.0260	0.0194	0.0170	0.0158	0.0132
C5	0.3250	0.0858	0.0946	0.0920	0.0712	0.0530	0.0490	0.0358
C6	0.6166	0.5396	0.6284	0.6286	0.5420	0.3198	0.2518	0.2392
C7	0.4598	0.2834	0.2284	0.2292	0.2272	0.2356	0.2264	0.2264
C8	0.4262	0.2956	0.2248	0.2250	0.2302	0.2340	0.2238	0.2240
C9	0.6904	0.6058	0.4334	0.4344	0.4188	0.4286	0.4142	0.4162
C10	0.5966	0.5966	0.4352	0.4296	0.4186	0.4286	0.4160	0.4162
Avg	0.3277	0.2667	0.2337	0.2332	0.2169	0.1804	0.1600	0.1575

Fig. 6.3 Test problem CEP testing results after ultrathin 8-layer deep learning enhancements

wide network versus thin network architecture and on other test problems will require further experiments.

6.10 Wide 2-Layer ARC Performance on the Theoretical Classification Problems

Here we compare the performance of a wide 2-Layer ARC network with the out of sample CEP testing scores of five well-known commercially available classification algorithms to determine where a wide 2-layer ARC network ranks in competitive comparison. The following table lists each classification algorithm in descending order of average CEP scores on all ten theoretical test problems. The lower the CEP the more accurate the classification results. The ARCN4 network is composed of 2 hidden layers each of which has 200 algebra neuron factories for a total of 400 algebra neuron factories in the entire network (width = 200, depth = 2) (Fig. 6.4).

The ARCN4 network has 5 times the total algebra neuron factories as the ARCN3 network although its depth is only 25% of the depth of the ARCN4 network. A comparison of their results highlights the advantages of a few wide layers over more thin layers. The top performer overall by a slight margin is now the ARCN4 wide 2-Layer algebra network. The Gradient Boosted Trees Learner (GBTL) has fallen further behind. Interestingly, a network of two wide hidden layers, with 200 algebra neuron factories per layer, for a total of 400 algebra neuron factories (200 algebra neuron factories per layer producing an average of 622.6 M concrete algebra neurons per test case) was enough to boost performance significantly beyond the ARCN3 thin 8-Layer algebra network. Whether this slight advantage persists for other wide network versus thin network architecture and on other test problems will require further experiments.

Test	MLP	DTL	TEL	RFL	ARC^{N1}	GBTL	ARC^{N2}	ARC^{N3}	ARC^{N4}
C1	0.0072	0.0724	0.0496	0.0492	0.0138	0.0308	0.0004	0.0000	0.0000
C2	0.0360	0.0740	0.0648	0.0664	0.0116	0.0240	0.0004	0.0016	0.0000
C3	0.0724	0.0972	0.1526	0.1522	0.0132	0.0332	0.0022	0.0026	0.0008
C4	0.0472	0.0174	0.0252	0.0260	0.0194	0.0170	0.0158	0.0132	0.0068
C5	0.3250	0.0858	0.0946	0.0920	0.0712	0.0530	0.0490	0.0358	0.0190
C6	0.6166	0.5396	0.6284	0.6286	0.5420	0.3198	0.2518	0.2392	0.2074
C7	0.4598	0.2834	0.2284	0.2292	0.2272	0.2356	0.2264	0.2264	0.2262
C8	0.4262	0.2956	0.2248	0.2250	0.2302	0.2340	0.2238	0.2240	0.2230
C9	0.6904	0.6058	0.4334	0.4344	0.4188	0.4286	0.4142	0.4162	0.4148
C10	0.5966	0.5966	0.4352	0.4296	0.4186	0.4286	0.4160	0.4162	0.4156
Avg	0.3277	0.2667	0.2337	0.2332	0.2169	0.1804	0.1600	0.1575	0.1513

Fig. 6.4 Test problem CEP testing results after wide 2-layer deep learning enhancements

6.11 Wide 8-Layer ARC Performance on the Theoretical Classification Problems

Here we compare the performance of a wide 8-Layer ARC network with the out of sample CEP testing scores of five well-known commercially available classification algorithms to determine where a wide 8-layer ARC network ranks in competitive comparison. The following table lists each classification algorithm in descending order of average CEP scores on all ten theoretical test problems. The lower the CEP the more accurate the classification results. The ARCN5 network is composed of 8 hidden layers each of which has 200 algebra neuron factories for a total of 1600 algebra neuron factories in the entire network (width = 200, depth = 8) (Fig. 6.5).

The top performer overall by a larger margin is now the ARCN5 wide 8-Layer algebra network. The Gradient Boosted Trees Learner (GBTL) has fallen still further behind. Interestingly, a network of eight wide hidden layers, with 200 algebra neuron factories per layer, for a total of 1600 algebra neuron factories (200 algebra neuron factories per layer producing an average of 1.5 B concrete algebra neurons per test case) was enough to boost performance significantly beyond the ARCN4 wide 2-Layer algebra network.

6.12 Conclusion

These experiments strongly indicate that there is a performance advantage when GLM algebraic expressions are fitted together in a feed forward acyclic network reminiscent of deep learning neural networks. However, these results are statistically indicative only. A great deal more research remains to be done.

There are other tools which use simple GP and/or GP merged with multiple regression, logit regression, and discriminant analysis. More experiments can help

Test	MLP	DTL	TEL	RFL	ARC^{N1}	GBTL	ARC^{N2}	ARC^{N3}	ARC^{N4}	ARC^{N5}
C1	0.0072	0.0724	0.0496	0.0492	0.0138	0.0308	0.0004	0.0000	0.0000	0.0000
C2	0.0360	0.0740	0.0648	0.0664	0.0116	0.0240	0.0004	0.0016	0.0000	0.0000
C3	0.0724	0.0972	0.1526	0.1522	0.0132	0.0332	0.0022	0.0026	0.0008	0.0000
C4	0.0472	0.0174	0.0252	0.0260	0.0194	0.0170	0.0158	0.0132	0.0068	0.0000
C5	0.3250	0.0858	0.0946	0.0920	0.0712	0.0530	0.0490	0.0358	0.0190	0.0000
C6	0.6166	0.5396	0.6284	0.6286	0.5420	0.3198	0.2518	0.2392	0.2074	0.2056
C7	0.4598	0.2834	0.2284	0.2292	0.2272	0.2356	0.2264	0.2264	0.2262	0.2044
C8	0.4262	0.2956	0.2248	0.2250	0.2302	0.2340	0.2238	0.2240	0.2230	0.2228
C9	0.6904	0.6058	0.4334	0.4344	0.4188	0.4286	0.4142	0.4162	0.4148	0.0178
C10	0.5966	0.5966	0.4352	0.4296	0.4186	0.4286	0.4160	0.4162	0.4156	0.4156
Avg	0.3277	0.2667	0.2337	0.2332	0.2169	0.1804	0.1600	0.1575	0.1513	0.1066

Fig. 6.5 Test problem CEP testing results after wide 8-layer deep learning enhancements

determine whether deep learning algebra networks benefit just the ARC tool; or, if other GP tools also benefit from deep learning algebra networks and whether these tools also benefit in a statistically similar manner?

These experiments were performed on a set of ten specific theoretical classification problems. Are deep learning algebra networks also beneficial with regression problems and in other real world problem domains?

These experiments involve only feed forward, acyclic, symmetric (all hidden layers are the same width) algebra networks. More experiments can help determine whether other types of deep learning algebra networks are more advantageous and in which problem domains.

The training method for these deep learning algebra networks involved optimizing the neurons in each layer against the dependent variable Y. This was done forwards, neuron by neuron, layer by layer, from the first hidden layer to the last hidden layer. Most deep learning neural nets are trained with quite different methods, wherein all coefficients in the network are adjusted backwards, from the dependent variable Y back to the first hidden layer. Is there a training method, for algebra networks, which works backwards on all coefficients in the network? Would this or some other training method provide statistically superior performance? A tremendous advantage of GP deep learning is that we have much greater visibility into what is actually going on within the network, since each of the network algebra neurons are human readable. In the future, much of the mystery surrounding deep learning networks may be clarified.

Clearly there is much work remaining in studying deep learning algebra networks—far more work than our single research group has resources. As we pursue our continuing studies of algebra networks, it is our hope that these experiments will excite other researchers to pursue the many questions still remaining with GP deep learning.

References

1. Augasta, M., Kathirvalavakumar, T.: Pruning algorithms of neural networks - a comparative study. Open Computer Sci. **3**(3), 105–115 (2013)
2. Berthold, M.R., Cebron, N., Dill, F., Gabriel, T.R., Kötter, T., Meinl, T., Ohl, P., Thiel, K., Wiswedel, B.: KNIME-the Konstanz Information Miner: Version 2.0 and beyond. ACM SIGKDD Explorations Newsletter **11**(1), 26–31 (2009)
3. Friedman, J.H.: Regularized discriminant analysis. J. Amer. Stat. Assoc. **84**(405), 165–175 (1989)
4. Karaboga, D., Akay, B.: A survey: algorithms simulating bee swarm intelligence. Artif. Intell. Rev. **31**(1–4), 61 (2009)
5. Korns, M.F.: A baseline symbolic regression algorithm. In: Genetic Programming Theory and Practice X. Springer, Berlin (2012)
6. Korns, M.F.: Predicting corporate forward 12 month earnings. In: Parpinelli, R., Lopes, H.S. (eds.) Theory and New Applications of Swarm Intelligence. Tech Academic Publishers, Cambridge (2012)
7. Korns, M.F.: Extreme accuracy in symbolic regression. In: Genetic Programming Theory and Practice XI, pp. 1–30. Springer, Berlin (2014)

8. Korns, M.F.: Extremely accurate symbolic regression for large feature problems. In: Genetic Programming Theory and Practice XII, pp. 109–131. Springer, Berlin (2015)
9. Korns, M.F.: Trading volatility using highly accurate symbolic regression. In: Handbook of Genetic Programming Applications, pp. 531–547. Springer, Berlin (2015)
10. Korns, M.F.: Highly accurate symbolic regression with noisy training data. In: Genetic Programming Theory and Practice XIII, pp. 91–115. Springer, Berlin (2016)
11. Korns, M.F.: An evolutionary algorithm for big data multiclass classification problems. In: Genetic Programming Theory and Practice XIV. Springer, Berlin (2017)
12. Korns, M.F., May, T.: Strong typing, swarm enhancement, and deep learning feature selection in the pursuit of symbolic regression-classification. In: Genetic Programming Theory and Practice XVI, pp. 59–84. Springer, Berlin (2019)
13. Koza, J.R.: Genetic Programming: On the Programming of Computers by Means of Natural Selection. MIT Press, Cambridge (1992)
14. McLachlan, G.: Discriminant Analysis and Statistical Pattern Recognition. Wiley, New York (2004)
15. Nelder, J., Wedderburn, R.W.: Generalized linear models. J. R. Stat. Soc. **135**, 370–384 (1972)
16. Peng, C.Y.J., Lee, K.L., Ingersoll, G.M.: An introduction to logistic regression analysis and reporting. J. Educ. Res. **96**(1), 3–14 (2002)
17. Platt, J.: Sequential minimal optimization: a fast algorithm for training support vector machines. Technical Report MSR-TR-98-14, Microsoft Research (1998)
18. Rout, A., Dash, P.K., Dash, R., Bisoi, R.: Forecasting financial time series using a lowcomplexity recurrent neural network and evolutionary learning approach. J. King Saud Univ.-Computer Inf. Sci. **29**, 536–552 (2017)
19. Schmidhuber, J.: Deep learning in neural networks: an overview. Neural Netw. **61**, 85–117 (2015)
20. Selvin, S., Vinayakumar, R., Gopalakrishnan, E., Menon, V., Soman, K.P.: Stock price prediction using LSTM, RNN and CNN-sliding window mode. In: International Conference on Advances in Computing, Communications and Informatics (ICACCI), pp. 1643–1647 (2017)
21. Tsantekidis, A., Passalis, N., Tefas, A., Kanniainen, J., Gabbouj, M., Iosifidis, A.: Forecasting stock prices from the limit order book using convolutional neural networks. In: 19th IEEE Conference onBusiness Informatics (CBI) 2017, pp. 7–12. IEEE Press (2017)
22. Uyanık, G.K., Güler, N.: A study on multiple linear regression analysis. Proc.-Soc. Behav. Sci. **106**, 234–240 (2013)
23. Vijh, M., Chandola, D., Tikkiwal, V., Kumar, A.: Stock closing price prediction using machine learning techniques. In: International Conference on Computational Intelligence and Data Science (ICCIDS 2019) (2019)

Chapter 7
Back to the Future—Revisiting OrdinalGP and Trustable Models After a Decade

Mark Kotanchek and Nathan Haut

Abstract OrdinalGP (2006) [4] embraced a fail-fast philosophy to efficiently model very large data sets. Recently, we realized that it was also effective against small data sets to reward model generalization. ESSENCE (2009) [6] extended the OrdinalGP concept to handle imbalanced data by using the SMITS algorithm to rank data records according to their information content to avoid locking into the behavior of heavily sampled data regions but had the disadvantage of computationally-intensive data conditioning with a corresponding fixed data ranking. With BalancedGP (2019) we shifted to a stochastic sampling to achieve a similar benefit. Trustable models (2007) [3] exploited the diversity of model forms developed during symbolic regression to define ensembles that feature both accurate prediction as well as detection of extrapolation into new regions of parameter space as well as changes in the underlying system. Although the deployed implementation has been effective, the diversity metric used was data-centric so alternatives have been explored to improve the robustness of ensemble definition. This chapter documents our latest thinking, realizations, and benefits of revisiting OrdinalGP and trustable models.

7.1 Introduction

Multi-objective symbolic regression rewarding both simplicity and accuracy has proven to be a game-changer for real-world data analysis due to the model development efficiency as well as the peripheral insights to be gained from analyzing a candidate pool of accurate-but-simple models—feature selection, metavariables, variable associations, etc [5].

M. Kotanchek
Evolved Analytics LLC, Rancho Santa Fe, CA, USA
e-mail: Mark@Evolved-Analytics.com

N. Haut (✉)
Michigan State University, Lansing, MI, USA
e-mail: hautnath@msu.edu

W. Banzhaf et al. (eds.), *Genetic Programming Theory and Practice XVIII*,
Genetic and Evolutionary Computation,
https://doi.org/10.1007/978-981-16-8113-4_7

In this chapter, we return to some of the seminal concepts explored in the early phases with the benefit of perspective gained over the past fifteen years. Most notably, we propose a new variant on **OrdinalGP** that offers benefits for the lumpy data sets often encountered in practice as well as revisit schemes for defining model ensembles to create trustable models.

These will be explored after we review some of the foundational concepts in the next section.

7.2 In the Beginning

7.2.1 Model Complexity—Getting What You Measure

With thousands of models being generated via classic Koza-style GP, we had the problem of determining which of the myriad possibilities should be deployed in industrial practice. Guido Smits realized that a good model was both accurate and simple. He also recognized that simple metrics like leaf counts, node counts, depth, etc. did not provide enough fidelity at the low end (which is our practical interest) so he synthesized the **ModelComplexity** metric. Originally stated as the sum of the node count of all possible subtrees, Maarten Keijzer subsequently recognized that this could also be expressed as the visitation length from the root node to all of the nodes as well as that it was a preferred computationally efficient metric for complexity [2]. It also has a bias towards bushier trees which opposes the tree-based crossover bias towards longer ones. Examples of the complexity metric can be seen in Fig. 7.1.

7.2.2 ParetoGP—Simplicity and Accuracy

Once we started thinking about models from a multi-objective viewpoint, the natural next step was to incorporate that perspective into the model development. The original **ParetoGP** algorithm (2004) [7] featured a punctuated equilibrium using single-objective tournament selection with preservation across cataclysms managed on a multi-objective criterion. Adoption of a multi-objective framework instantly sped up the model search by a factor of 60 while simultaneously eliminating bloat as a problem—it has not reemerged over the course of the past fifteen-plus years.

There are two key assumptions that are implicit to **ParetoGP**:

- only a relative few number of variables drive the observed behavior and
- simple and accurate models are most desirable (in other words, we do not want to chase R^2 irrespective of complexity).

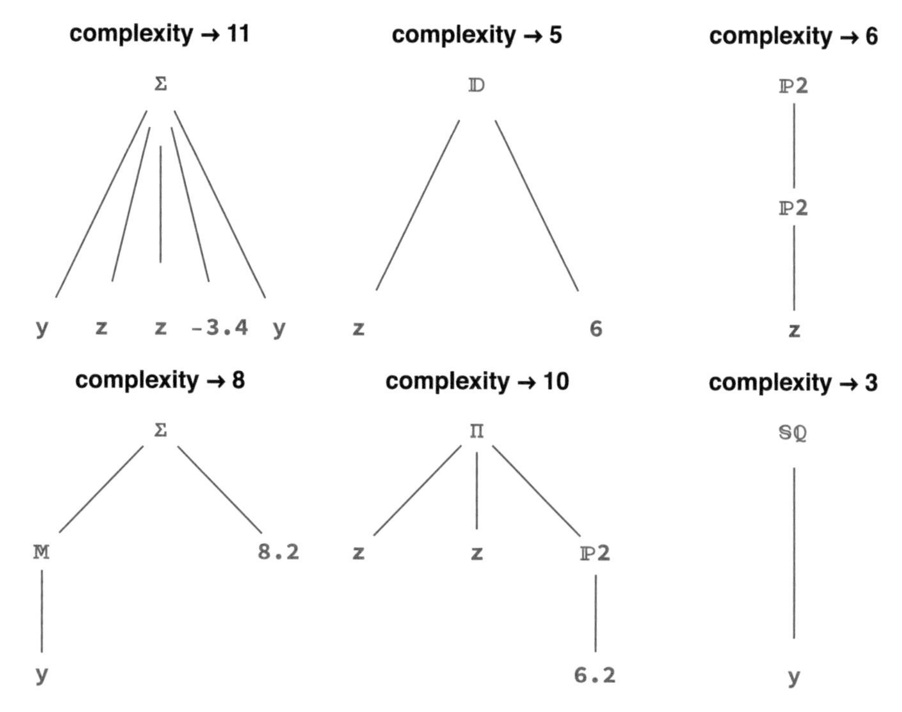

Fig. 7.1 The **ModelComplexity** is most easily visualized as the *visitation length*—i.e., number of nodes traveled through from the root node to all possible nodes (including the root node)

7.2.2.1 ParetoTourney

Tournaments are the preferred choice in single-objective evolutionary computing because they are robust and easily tunable with respect to the focusing from the candidate pool. With the development of the **ParetoTournament** (2006) [4], we were able to easily migrate classic GP to a multi-target implementation. As we see in Fig. 7.2, the evolutionary energy is focused on the models that best balance the competing objectives. (To avoid losing useful genetics, we also define a multi-objective Methuselah set that also get a free transfer into the subsequent generation.)

7.2.3 Secondary and Alternating Objectives

Age Layered GP is a useful concept to drive continuous innovation so we added **ModelAge** as a secondary modeling objective (used during development but not returned to the user). Other criteria such as **ModelDimensionality** or **BasisSet-Count** or **ModelNonlinearity** are also potentially useful to guide model development. Alas, the curse-of-dimensionality comes into play in that the focusing power

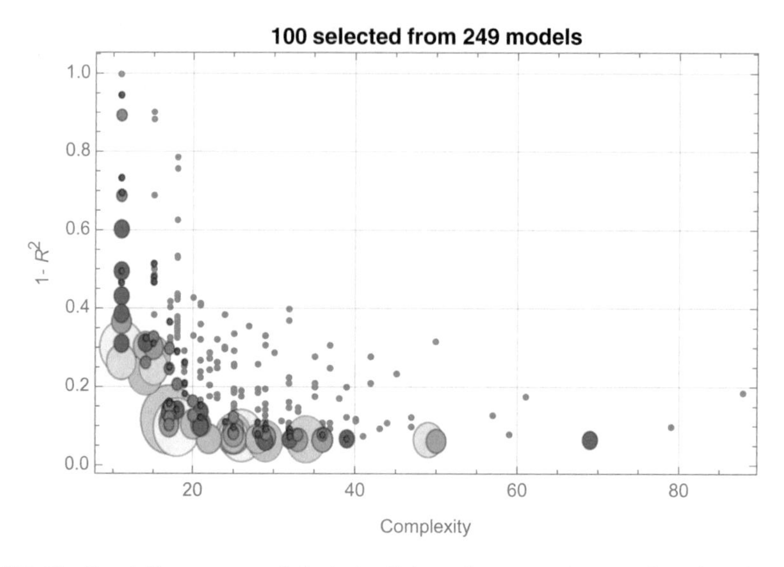

Fig. 7.2 The **ParetoTourney** rewards both simplicity and accuracy; however, it tends to focus the evolutionary effort at the knee of the **ParetoFront**—which is of most practical interest and why bloat is eliminated as a concern. Here the candidate population was randomly sampled in groups of 30 and the **ParetoFront** models from the tourneys accumulated until the desired next generation breeders (300 models) were designated. The size of the bubbles is proportional to the frequency of being selected. In this case, 100 of the candidates received breeding rights, albeit, not equally

of **ParetoTournaments** quickly becomes diffused as additional criteria are added. The approach we use to address this is to use alternating objectives (2005) [7] and use a different subset of the desired criteria with each generation. This is directly analogous to Epsilon-Lexicase Selection (2016) [1].

7.2.4 OrdinalGP—Failing Forward

Evaluating a crappy model against a large data set takes the same amount of time as evaluating a good model. However, if our goal is simply to partition candidate models into a keep/discard classification, we do not need to know the precise performance. Instead, if we evaluate against a data subset we can avoid a significant computational load. Towards this end, **OrdinalGP** (2006) [4] chooses a random data subset for each generation. In addition to the computational gains and additional model refinement achieved due to more generations of development, there is an additional benefit since the stochastic nature of the fitness landscape rewards model generality.

To make the benefits tangible, assume that evaluating a model against a million records requires 1 s of CPU time. If we have a population of 300, then five minutes are required per generation. Chopping the generational assessment down to 1% (aka,

10,000 records) implies that we can get through many more generations of algorithmic development. Since there is rarely a million records worth of information in the data, the operational gains can be substantial. Of course, for comparison purposes, we do need to evaluate the final models against the complete data set.

Think of this as K-fold cross-validation on steroids.

7.2.5 Ensembles—Trustable Models and Active Design-of-Experiments

The focus on simplicity meant that feature selection was facilitated and the user could drive modeling towards causal inputs which were most controllable. The problem of picking out THE model from the myriad contenders (which, eventually, led to the **ParetoGP** approach) remained. Rather than picking THE model, we realized that a **ModelEnsemble** (2007) [3] of diverse models from near the knee of the **ParetoFront** was, effectively, a trustable model that could detect extrapolation or other operating regime changes.

In addition to deployment trust, an obvious implication is that we can use ensembles to guide future data collection. Effectively, using the ensemble divergence identifies the most useful experiment to drive uncertainty out of the model.

7.2.6 Data Balancing

As practitioners, we are often faced with lumpy data rather than the balanced data from designed experiments. As such, the information content of the data records is not uniform. The SMITS data balancing and associated ESSENCE model search algorithms (2009) [6] addressed this by estimating the relative information content of the data records and building foundational global models which were enhanced in subsequent generations by the incremental inclusion of additional information.

Although a good idea, there were a number of practical issues associated with the identification of a proper information/distance/clustering metric in high-dimensional spaces as well as scaling issues to the really large data sets where it is most needed. However, the **BalancedGP** approach discussed later gets us most of the way to the behavior we want.

7.3 BalancedGP

The basic notion of **BalancedGP** is to meld **OrdinalGP** with data balancing concepts in a computationally efficient manner. The foundation is the **BalancedSample** function which divides the data into equal-increment response bins from which data records are stochastically pulled for each evolutionary generation.

7.3.1 *DataSubsetSize*

OrdinalGP demands a decision about the **DataSubsetSize** to be used in each generation. Intuitively, we would contend that for small data sets, we want to use most of the data each generation (the limit being leave-one-out-cross-validation) while for very large sets a small fraction should be sufficient.

As we saw in Fig. 7.3, not every data record has equal information content so we have the question as to how much data is needed to capture the essence of the underlying data? Our approach was simply to synthesize a data table—100% of the data at 100 points, 25% at 1,000 points, 12.5% at 10,000 points, etc. until we reach 1% at a million records—and to use SymbolicRegression to evolve a reasonable approximation of the percent of data records to be used. The chosen expression, $33 + 8215/(5.3 + numRecords) - 2.3Log[numRecords]$ is clipped at 100% for data set sizes below 100 and 1% beyond a million records. Although somewhat ad hoc, using **OrdinalGP** for our test suite of real-world data shows that such a profile beats **ClassicGP** with the full data set evaluation. Figure 7.4 shows the relationship between the total number of data records and data subset size.

7.3.2 *BalancedSample*

Reality is that we often have essentially the same data point repeated due to closed-loop control in process systems or the cellular equivalent for biological systems. If our goal is to develop a global model, we wish to de-weight such regions so the more sparsely observed behavior can emerge during the model development. **OrdinalGP** doesn't really address this since, even though the samples change with each generation, the distribution will remain comparable—albeit stochastic.

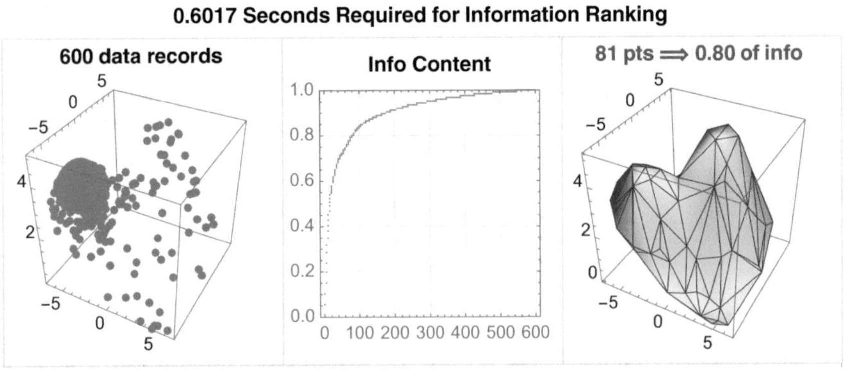

Fig. 7.3 Not all information is equally informative. Choosing the right data subset can convert an imbalanced data set into a balanced one—which is easier to use for model development

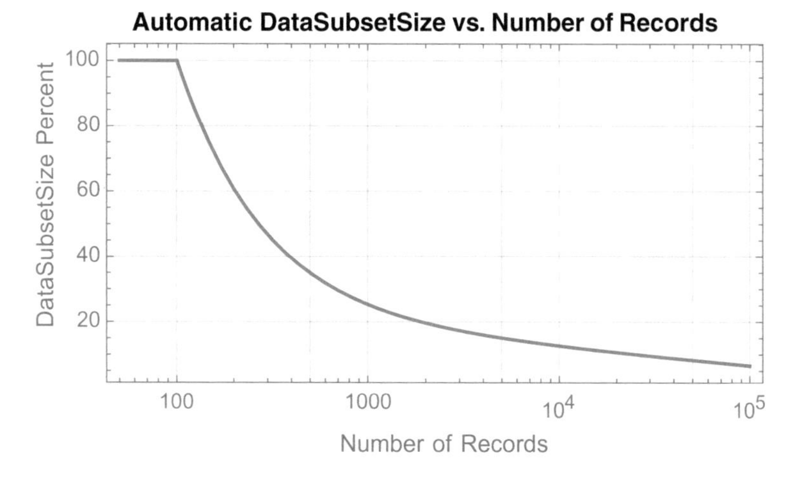

Fig. 7.4 The default percentage of data used during **OrdinalGP** and **BalancedGP** changes as a function of data set size. Despite being an ad hoc formula, it works well in practice

Algorithms for deciding the number of range bins

DataSegment Comparison

# Records	Sturges	Rice Rule
10	5	5
100	8	10
1000	11	20
10 000	15	44
100 000	18	93
1 000 000	21	200

DataSegment Schemes

Number of Bins: 80, 60, 40, 20

Rice→ $\left\lceil 2 \sqrt[3]{n} \right\rceil$

Sturges → $\left\lceil \frac{\log(n)}{\log(2)} \right\rceil + 1$

Number of Numeric Samples: 1, 10, 100, 1000, 10^4, 10^5

Fig. 7.5 It seemed reasonable to leverage the histogramming rules to automatically determine the number of response range bins for **BalancedGP**. Of the two most popular, we prefer the Rice rule

The **BalancedSample** approach during **BalancedGP** is to partition the data according to equal increments in the target response with the result that the bins have different numbers of data records. For each generation, the bins are equally sampled to achieve the desired number of records with any shortfalls made up by randomly sampling from the overall set. To determine the number of response range bins, we stole from the histogramming community and settled in on the Rice rule (Fig. 7.5).

Additional computational efficiencies are possible if we preprocess the data set by sorting by the response and specify index ranges. However, even without that,

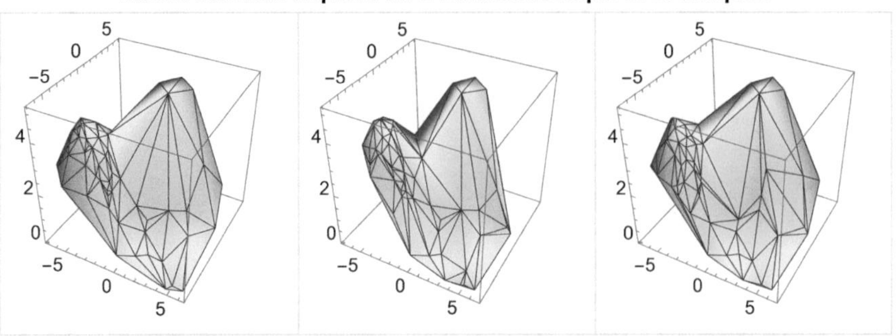

0.0052 seconds required for a BalancedSample of 81 samples

Fig. 7.6 For clarity, we have reduced the number of samples from the default 192 to that used above for the SMITS sampling. Note in this case that the sampling changes across generations which helps to enforce generalization. Three sample sets are shown

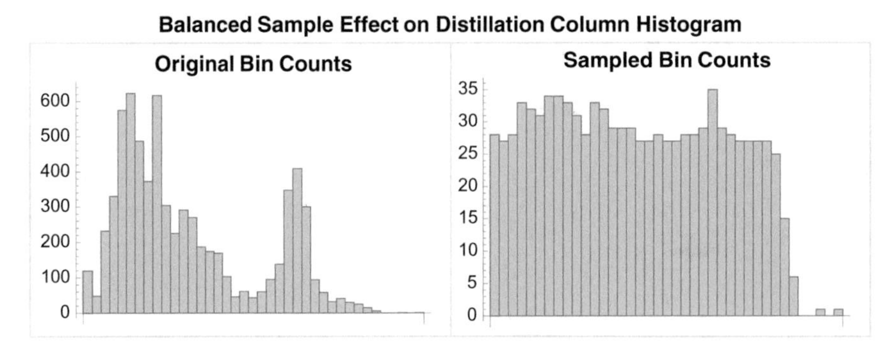

Fig. 7.7 The 7,000 data records in the distillation tower data set are not evenly distributed to to coming from an production system operating in closed-loop control. The default for this size of data set would be to sample 960 data records—which does a pretty good job of flattening the lumpiness in the data. **OrdinalGP** would follow the original distribution

the computational benefits outweigh those of the SMITS algorithm's information content assessment. Figure 7.6 shows three separate sample sets which, clearly, are comparable to that from the SMITS algorithm.

Although we could balance across all of the variables used in model development, for efficiency reasons, we just balance the response during **SymbolicRegression**. The implications of this are illustrated in Fig. 7.7. One side-effect of this algorithm is that data bins with few members may be fully represented in every generation—which could expose an outlier influence risk.

7.3.3 BalancedGP

At the end of **SymbolicRegression** if **BalancedGP** or **OrdinalGP** is used, the models are evaluated against the full data set. In Fig. 7.8 we look at the results from short model searches which illustrate the benefits of being able to grind through more generations.

Although the **BalancedGP** and **OrdinalGP** are fairly comparable from a global metric (both having been evaluated against the full data set), anecdotally, **BalancedGP** seems to handle edge conditions a little better. Spurious correlations can impose a lower bound on the size of the data set which benefits from the application of **OrdinalGP** or **BalancedGP**.

BalancedGP has an intrinsic advantage if our targeted response is categorical and imbalanced since it will automatically balance the competing categories which helps towards our implicit goal of a global model.

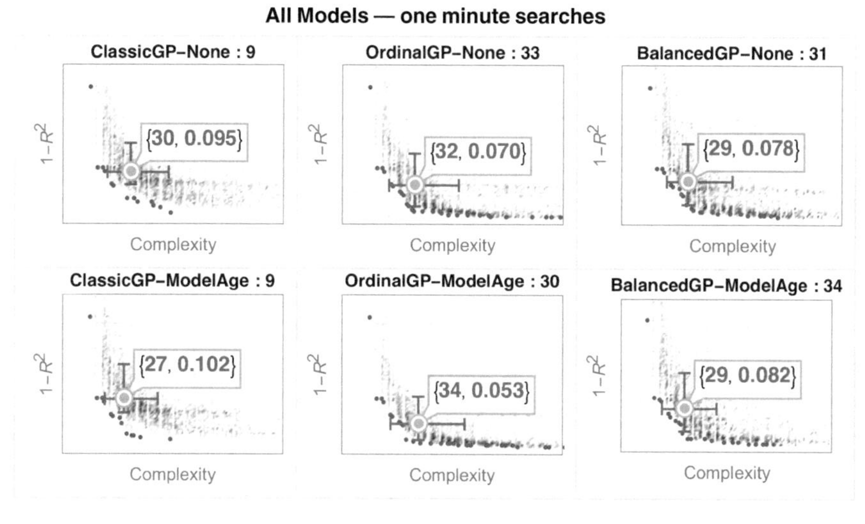

Fig. 7.8 Here we compare **ClassicGP** against **OrdinalGP** and **BalancedGP** for one-minute model searches with and without using **ModelAge** as a secondary search criteria. 32 **IndependentEvolutions** were run for each combination and the average maximum **ModelAge** shown in the plot title. The **OrdinalGP** and **BalancedGP** get through three times as many generations so there is considerably more model refinement. For these relatively short model searches, it does not appear that the innovation preservation afforded by using **ModelAge** as a secondary criteria provides a noticeable benefit

7.4 Ensembles

7.4.1 Introduction to Ensembles

The trustability factor of a diverse model ensemble provides a huge practical advantage—knowing that you are pushing a chemical plant into previously unexplored operating regimes is significant! For most ensembles, we choose from a candidate set of simple-and-accurate using a common handful of variables. (Ensembles with diverse variable sets can be used to detect sensor failure; however, we will not address that nuance herein.)

In Fig. 7.9 we isolate on distillation column models which only contain the most popular 4-variable combination and build an ensemble two ways:

- *Uncorrelated Models*: the strategy is to build a correlation matrix of the model residuals and look for the least-correlated pair and, from that foundation, look for the model which is least correlated to that pair and continue until the desired number of models is achieved.

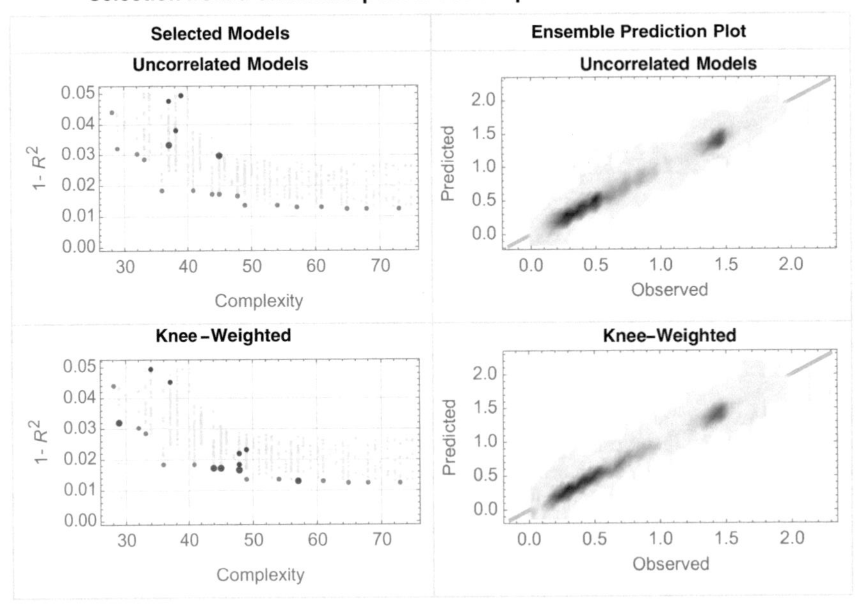

Fig. 7.9 Here we have identified the five least correlated models (in terms of their residuals) as well as a knee-weighted selection. The key takeaway is that all of the candidate models are of high quality so the resulting predictions are still robust

Uncorrelated Models

	Complexity	$1-R^2$	Function
1	37	0.033	$3.38-\frac{233.26}{\text{refluxFlow}}-(9.27\times10^{-4})\,\text{refluxFlow colTemp}_1+(1.02\times10^{-5})\,\text{feedFlow colTemp}_1{}^2$
2	37	0.047	$-0.34+\frac{(5.33\times10^{-5})\,\text{feedFlow}^4\,\text{colTemp}_1{}^2}{\text{refluxFlow}^4}$
3	38	0.038	$1.17-(9.91\times10^{-3})\,\text{refluxFlow}-(2.76\times10^{-5})\,\text{colTemp}_1{}^3+(2.49\times10^{-7})\,\text{colTemp}_1{}^3\,\text{upstreamFlow}_2$
4	39	0.049	$0.16+(5.97\times10^{-14})\,\text{feedFlow}\,(-\text{feedFlow}+\text{refluxFlow})^2\,\text{colTemp}_1{}^4$
5	45	0.030	$-8.20+\frac{87.07}{\text{colTemp}_1}+0.27\,\text{colTemp}_1+\frac{38.61}{\text{upstreamFlow}_2}-\frac{0.13\,\text{refluxFlow colTemp}_1}{\text{upstreamFlow}_2}$

Fig. 7.10 The **UncorrelatedModels** selected in Fig. 7.9 are diverse in structure despite being comparable in terms of complexity and accuracy

- *Knee-Weighted*: The candidate models are partitioned into four groups:
  - candidates
  - **ParetoFront** of candidates
  - knee models of candidates (better than the median value of quality criteria)
  - **ParetoFront** of knee models

  From each of the partitions, we identified the least correlated models as well as the most-typical (largest contribution to the dominant eigenvalue) and remove any duplicates. This generally results in 8–10 models being selected. The goal here is to over-weight the knee of the **ParetoFront** and try to include other models to detect extrapolation.

If we have a large number of candidate models (as in this case) which would imply a very large correlation matrix and a correspondingly large number of correlation computations, the models are randomly partitioned into sets of less than 100 with the least correlated selected from each, results merged and the process continued until the number of models being considered is less than the threshold. In this case, the selected model set will be stochastic. The diversity of selected model forms is illustrated in Fig. 7.10. Figure 7.11 shows the response behavior of the two ensembles and demonstrates the ability to detect extrapolation into new regions of parameter space—which is a major advantage of ensembles relative to choosing THE model.

7.4.2 Ensembles of the Future

Any mechanism to identify diverse models for inclusion into an ensemble will provide many of the benefits of a trustable model. However, we have a few possible sins of omission which might offer enhancements.

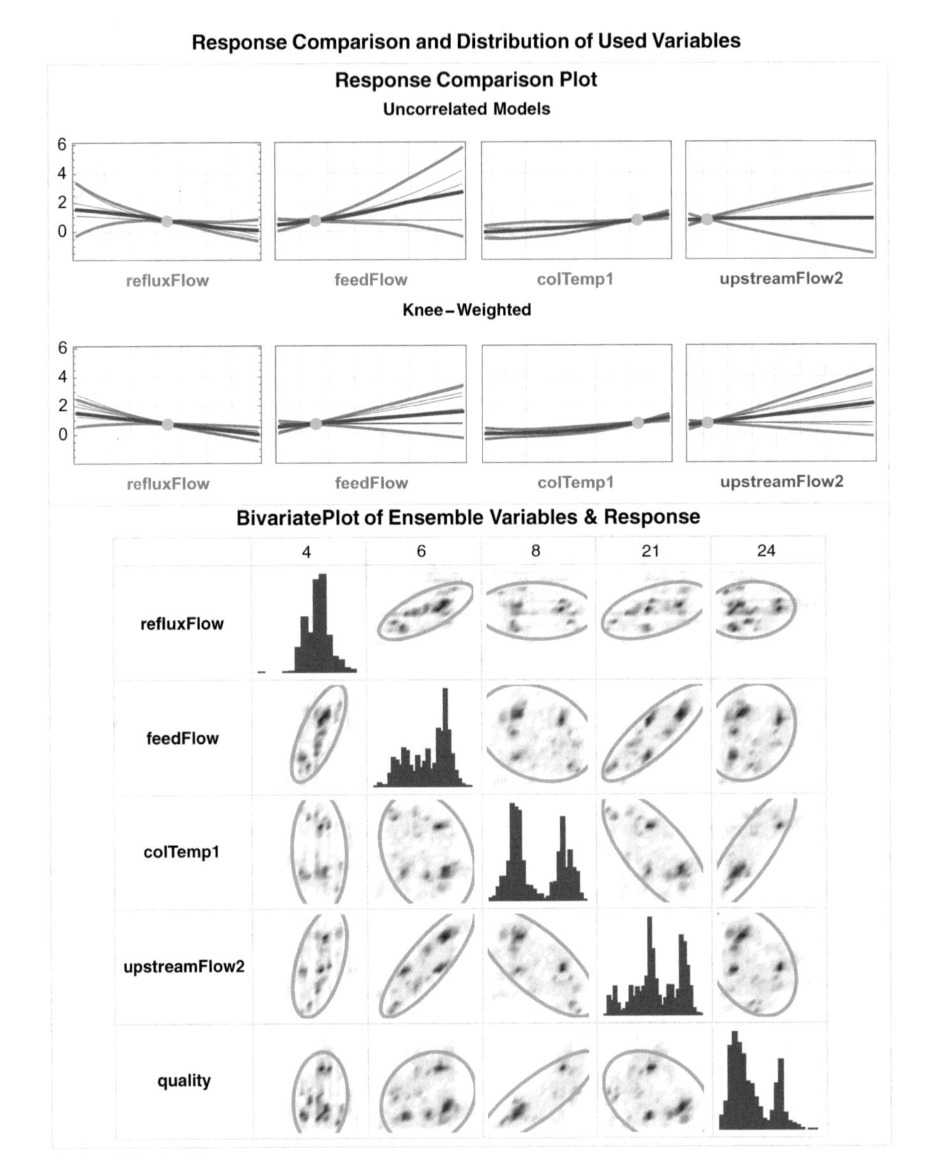

Fig. 7.11 The green dot in the **ResponseComparisonPlot** represents the reference data point at which the model is evaluated and, in this case, is one of the data records. The plots show the response behavior (blue line—aka, ensemble median) if we change a variable while holding the other values at their reference point. The gray lines show the trajectories of the constituent models. The yellow envelope shows the 2σ boundaries of the constituent models. From the above, we see that the models agree where constrained by data but diverge if asked to extrapolate into new regions of parameter space

7.4.2.1 Global Versus Greedy Model Selection

The shipping algorithm is greedy but efficient in that it looks for the least coupled models and adds to those. We also looked at searching the correlation matrix for the globally least coupled model set as well as clustering the models and randomly selecting from the clusters. To some extent, however, large numbers of candidate models force the use of heuristics for computational efficiency. The conclusion was that although a global approach could provide a less correlated overall set of models, the runtime scaling of the global approach makes it very quickly unfeasible. As well, the ensemble quality is only very slightly improved in regards to extrapolation detection.

7.4.2.2 Data Coverage and Synthesized Data

Using the residual as a diversity metric has the fundamental problem that such may not cover the true operating space. Using synthesized data addresses this at the cost of not having a truth reference so lack of correlation becomes relative. Additionally, matching the true behavior of coupled and correlated inputs can be difficult since both dithering around observed points as well as learning joint distributions can be difficult. The conclusion was that the synthetic data approaches worked well when realistic data was easy to generate but deteriorated quickly as the quality of synthetic data decreased. Further research to improve the quality of synthetic data could bring more appeal to using synthetic data as a guide for ensemble development.

7.4.2.3 Data Balancing

Currently, we use the entire data set to characterize diversity. However, the data balancing explored in association with **BalancedGP** inspires asking whether such should be used instead for the characterization. Such is desirable for efficiency reasons when building ensembles against large data sets as well as the lumpy ones. Additionally, it may be worthwhile to partition data into regions and select models based upon their performance in those regions. The initial exploration using Mathematica's built-in clustering functions to partition the data and selecting the best models from each data cluster to build the ensemble showed promising results. When the data was able to be partitioned effectively by Mathematica's clustering functions, the ensembles fit the data well and were able to detect extrapolation. Occasionally, the clustering function would create some data clusters containing outliers or very few points, so selecting models that fit these clusters decreased the quality of the ensembles. Further research to improve the selection of the data clusters, such as combining small clusters or eliminating outliers before partitioning the data, could eliminate the current issues with this approach.

7.5 Conclusions

In this chapter we have dusted off concepts and ideas first explored close to two decades prior and found that further exploration provided benefit for symbolic regression. Some like **BalancedGP** and **BalancedSample** have been integrated into *DataModeler* while others like strategies for ensemble definition continue to be explored.

In any event, symbolic regression is not a solved problem and remains rich in possibilities.

References

1. Cava, W. L., Spector, L., Danai, K.: Epsilon-lexicase selection for regression. In: Proceedings of the Genetic and Evolutionary Computation Conference (2016)
2. Keijzer, M., Foster, J.: Crossover bias in genetic programming. In: Genetic Programming, pp. 33–44. 10th European Conference, EuroGP (2007)
3. Kotanchek, M., Smits, G., Vladislavleva, E.: Exploiting trustable models via pareto GP For targeted data collection. In: Genetic Programming Theory and Practice VI, pp. 145–162. Springer, New York (2009)
4. Kotanchek, M., Smits, G., Vladislavleva, E.: Pursuing the pareto paradigm: tournaments, algorithm variations, and ordinal optimization. In: Genetic Programming Theory and Practice IV, pp. 167–185. Springer, New York (2007)
5. Kotanchek, M., Vladislavleva, E., Smits, G.: Symbolic regression is not enough: it takes a village to raise a model. In: Genetic Programming Theory and Practice X, pp. 187–203. Springer, New York (2013)
6. Kotanchek, M., Vladislavleva, E., Smits, G.: Symbolic regression via genetic programming as a discovery engine: insights on outliers and prototypes. In: Genetic Programming Theory and Practice VII, pp. 55–72. Springer, New York (2010)
7. Smits, G., Kotanchek, M.: Pareto-front exploitation in symbolic regression. In: Genetic Programming Theory and Practice II, pp. 283–299. Springer, New York (2005)

Chapter 8
Fitness First

W. B. Langdon

Abstract With side effect free terminals and functions it is possible to evaluate the fitness of genetic programming trees from their parents without creating them. This allows selection before forming the next generation. Thus avoiding unfit runt Genetic Algorithm individuals, which will themselves have no children. In highly diverse GA populations with strong selection, more than 50% of children need not be created. Even with two parent crossover, in converged populations, $e^{-2} = 13.5\%$ can be saved. Eliminating bachelors and spinsters and extracting the smaller genetic material of each mating before crossover, reduces storage in an N multi-threaded implementation for a population M to $\leq$0.63M+N, compared to the usual M+2N. Memory efficient crossover achieves 692 billion GP operations per second, 692 giga GPops, at runtime on a 16 core 3.8 GHz desktop.

8.1 Introduction

It is commonly held that genetic programming run time is dominated by the time to evaluate evolved individual program's fitness [7, 30]. However, in the last couple of years fitness evaluation for floating point problems has progressed enormously [3, 10, 11, 14, 15, 17], meaning in large programs of tens of millions of opcodes the primary cost can be in performing crossover rather than fitness evaluation, see Figs. 8.1, 8.2 and 8.3. We show the cost of subtree crossover can be reduced by (1) doing crossover after fitness and (2) separating the subtree donating parent (the dad). See Figs. 8.4, 8.5 8.10 and 8.11.

The next section summarises recent use of high performance parallel computing for tree based genetic programming. This is followed by Sect. 8.3 which describes how it is possible to assign fitness values to the current generation before it is complete by incrementally evaluating [15] children using only the crossover points and their

W. B. Langdon (✉)
Department of Computer Science, University College London, Gower Street, London WC1E 6BT, UK
e-mail: W.Langdon@cs.ucl.ac.uk

W. Banzhaf et al. (eds.), *Genetic Programming Theory and Practice XVIII*, Genetic and Evolutionary Computation,
https://doi.org/10.1007/978-981-16-8113-4_8

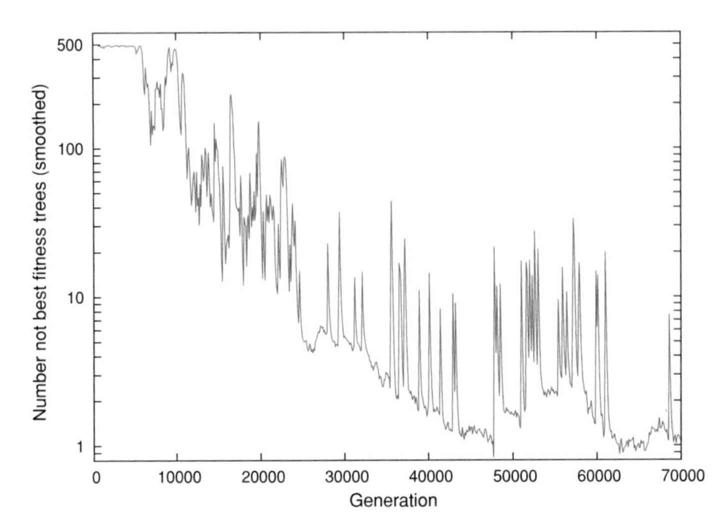

Fig. 8.1 Evolution of fitness convergence. Plot of number of individuals worse than the best smoothed by plotting running mean of 100 generations. Sudden upticks as new better individual is found and takes over the population. Pop = 500

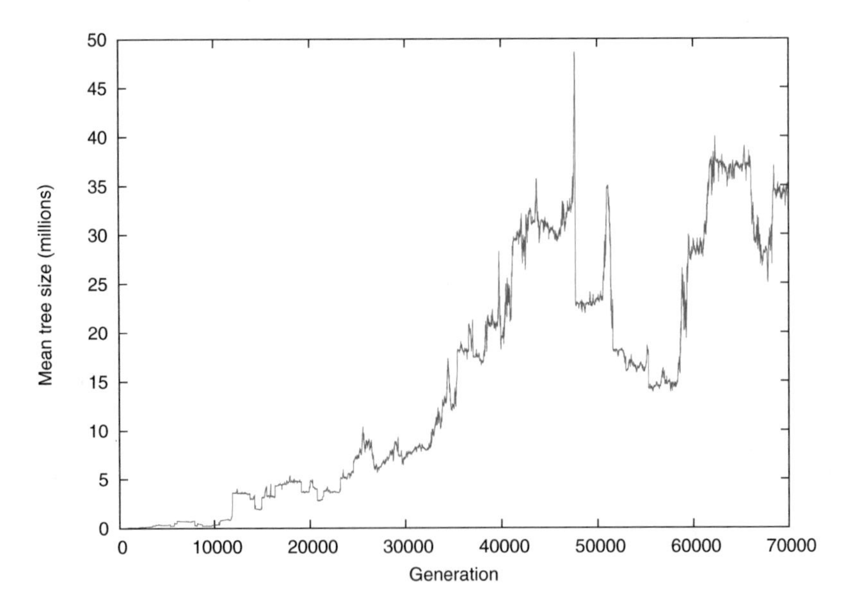

Fig. 8.2 Evolution of tree size

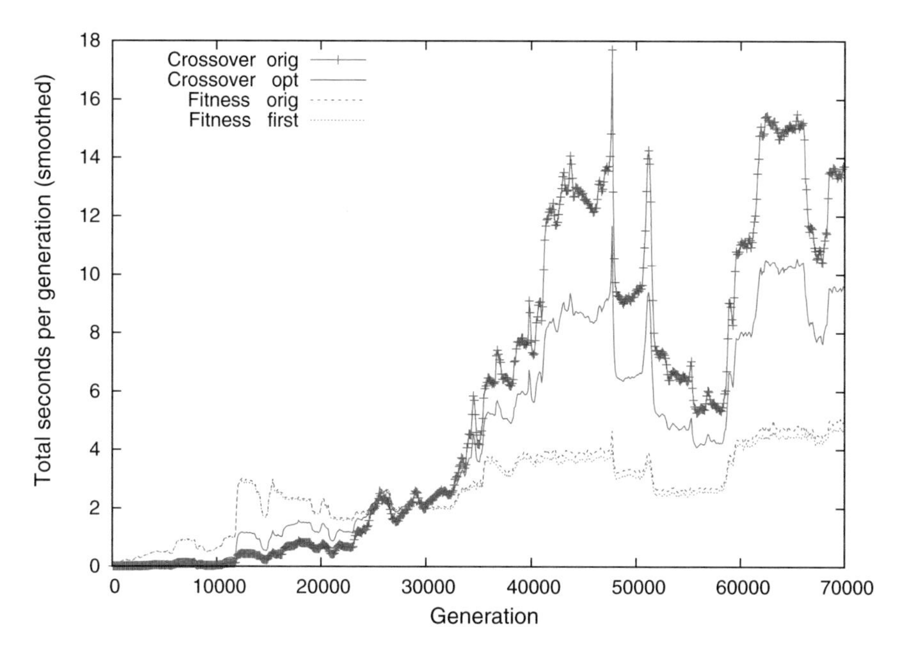

Fig. 8.3 Evolution of average time taken by incremental fitness first and crossover evaluation. Pop = 500. 16 core Intel 3.80 GHz i7-9800X. Running means of 100 generations

parents. Section 8.4 shows reversing the order of fitness and crossover allows us to avoid using crossover to create poor fitness individuals. Also separating subtrees from fathers eases other crossover optimisations, Sect. 8.4.1.

The final Sects. 8.5–8.8 deal with implementation issues and analysis. Section 8.5 says that, contrary to internet wisdom, current implementations of C++ memmove are not slow compared to memcpy and discusses its implications for our inplace crossover optimisation. Section 8.6 describes the GP's speed. Section 8.7 gives a brief model of the impact of tournament selection on diverse populations (such as those typically found near the start of GP runs). This complements the mathematical analysis in Sect. 8.3, which covers converged populations, when everyone has the same fitness. The two cases each have benefits which our crossover optimisations are able to exploit, leading to speedups both at the start and end of GP runs. Section 8.8 describes problems of load balancing to get peak performance from modern multi-core Intel CPUs before we conclude in Sect. 8.9. First we describe recent developments with speeding fitness evaluation and crossover using parallel hardware.

8.2 Faster Genetic Programming via Parallel Hardware

8.2.1 Multiple CPU Cores

Koza [7] described genetic programming as being embarrassingly parallel, in that by distributing the population, GP can easily be coded to get near 100% loading of parallel computers. Typically the population is spread across multiple computers which operate more or less independently. Similarly, our GP experiments are run on a parallel Intel multi-core desktop. There is a single administration thread, but with the creation of each individual in the population by crossover and also its incremental fitness evaluation being treated as separate tasks. These tasks are run in parallel by the hardware cores. The Linux posix pthreads environment is used with one thread per CPU core. Load balancing across the cores is achieved by each thread taking the next individual to be processed as it finishes the last, until the whole population has had its fitness calculated or the required members of the next population have been created using crossover.

This multi-threading strategy works well when the population size is much more than the number of CPU cores and the tasks are more or less the same size (but see Sect. 8.8.1) and means the population remains united. This approach also allows a light central core containing all the stochastic code with only resource intensive (deterministic) code running in parallel threads. Thus, with careful control of pseudo random number seeds, it makes it possible to replicate runs exactly in serial and different parallel environments. That is, a sequential run will produce the same sequence of populations as one using 8-cores, which in turn is the same as that produced on a 16-core machine. Indeed the system has been run on cluster nodes with 48 cores.

Note a single united panmictic population may converge more rapidly than in parallelisation schemes which require the population to be geographically divided between physically distinct processors. The next section considers a much finer grained parallelism in which fitness evaluation of a single individual is spread over up to 16 compute elements.

8.2.2 Multiple Fitness Cases Simultaneously

Our use of Intel's SIMD AVX-512 parallel vector instructions allows 16 test cases to be evaluated simultaneously [10, 11, 16]. This can be thought of as the floating point equivalent of Poli's sub-machine code GP [28]. With sub-machine code GP an opcode (e.g. AND) can be evaluated on 64 Boolean test cases at each clock tick [9, 27]. Indeed older AVX instructions have been used to evaluate 128 and 256 Boolean test cases simultaneously [6]. Also the newer AVX-512 instructions could be used to extend this to 512 test cases in parallel. Indeed genetic improvement

([18, 22–25, 32]) has been applied to AVX code itself [11]. Our latest developments [15] mean in extended GP runs the primary cost is creating and storing the next generation, rather than calculating its fitness.

8.2.3 Fitness First

It is relatively straight forward to convert our bottom up incremental evaluation [15] from evaluating each child directly, to evaluating it indirectly via its parents, Fig. 8.4. Thus we can find a child's fitness before creating the child. Figure 8.6 shows an example of incremental fitness evaluation using only the child's parents. Figure 8.7 shows an example from generation 1000 where incremental evaluation proceeds approximately half way from the crossover point to the root node. If it turns out the child is never used, e.g. because it is unfit or unlucky, it need not be created (Fig. 8.5).

We assume the GP population is made of pure functions (i.e. there are no side effects) and the same test cases are used to assign fitness of the children as were used to find the fitness of their parents.

Fitness first starts by evaluating the subtree to be removed from the mum (white) and the subtree to be inserted (black), Fig. 8.6. Apart from starting at the root of a subtree (i.e. within a parent) rather than at the root node, the evaluation is the same as usual. I.e., the normal depth first recursive evaluation is used for all subtrees that have to be evaluated. (Albeit if AVX-512 is supported in hardware, we use parallel AVX instructions.)

If, for all test cases, the values produced by the new code to be inserted are identical to those produced by the code to be removed, the inserted code has no effect and the child's fitness must be the same as the mum's. If any are different, we proceed up the mum tree towards its root.

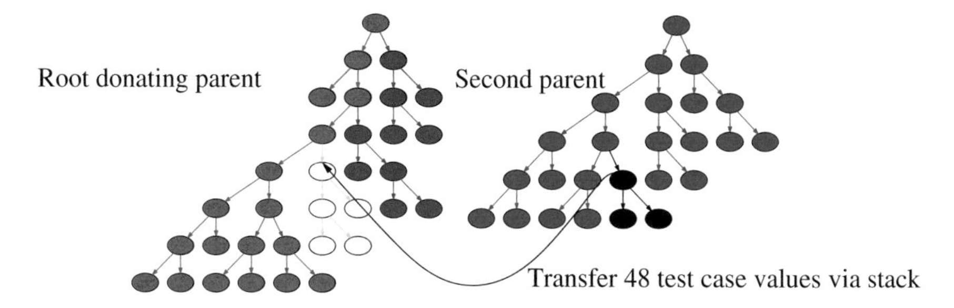

Fig. 8.4 Fitness is evaluated using only parents, i.e., before the child is created by crossover. Assuming no side effects, the subtree to be inserted (black) is evaluated on all test cases and values are transferred to evaluation of mum (left) at the location of the subtree to be removed (white). We use our incremental evaluation [15], so differences between original code (white subtree) and new are propagated up 1st parent (mum) until either all differences are zero or we reach the root node

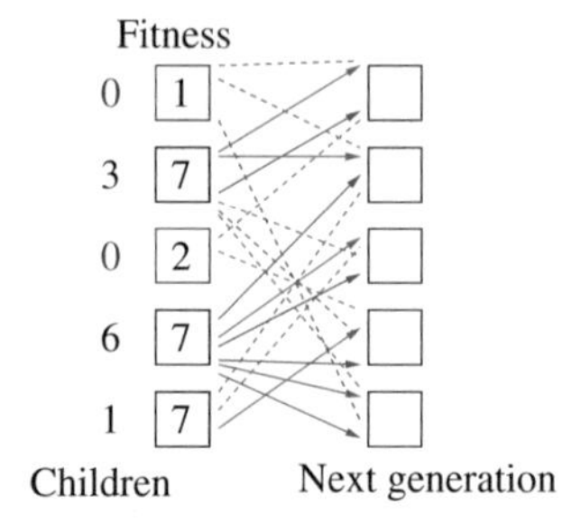

Fig. 8.5 As fitness can be calculated before crossover (Fig. 8.4), the parents can be chosen before crossover too. Here two low fitness individuals (fitness 1 and 2) have no children and hence their creation need not be completed. Lines indicate the two members of each tournament used to select the first (red) and second (blue) parent. Solid lines with arrows are the winners of each tournament [29]. (Binary tournament only for illustration, we actually use tournaments with 7 members.) All common EA selection schemes (with either mutation or crossover) are guaranteed to have members of the current population who will not have children in the next generation

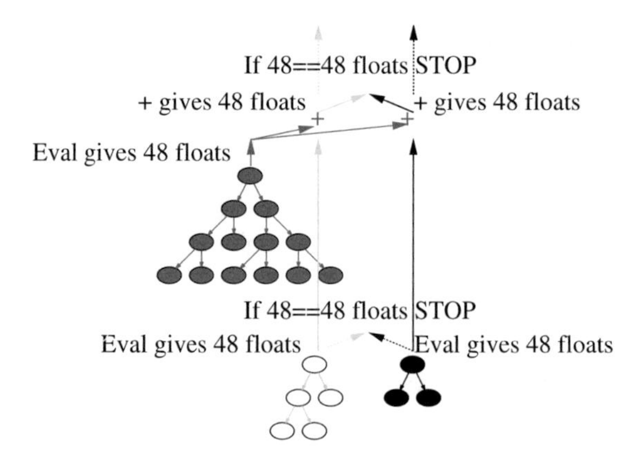

Fig. 8.6 "Fitness first" begins by evaluating the subtree to be removed from the mum (white) and the subtree to be inserted (black). It proceeds up the mum's tree until either the evaluation in the mum and unborn child are the same or it reaches the root node. The red subtree is in the mum but it is identical to the code in its child and so need be evaluated only once per test case. Note the code from the parents is evaluated without creating the child. Example from Fig. 8.4. See also Fig. 8.7

The example in Fig. 8.6 shows the next node up is a plus. We find the other subtree in the mum that is the plus' other argument (shown in red) and recursively evaluate it for all the test cases. Again this GP code (which must be identical to that in the as yet unborn child) is run in the mum in situ. The evaluation again gives a vector of floats (one element per test case). Next the function (plus) is applied to each value in the vector (red arrow) and the corresponding value from the mum subtree (light blue arrow) and similarly to the values from child's subtree (black arrow). This gives us two float vectors (one for mum and one for the child). Again if they are equal we can stop, since, if they are equal, they would remain equal all the way to the root node.

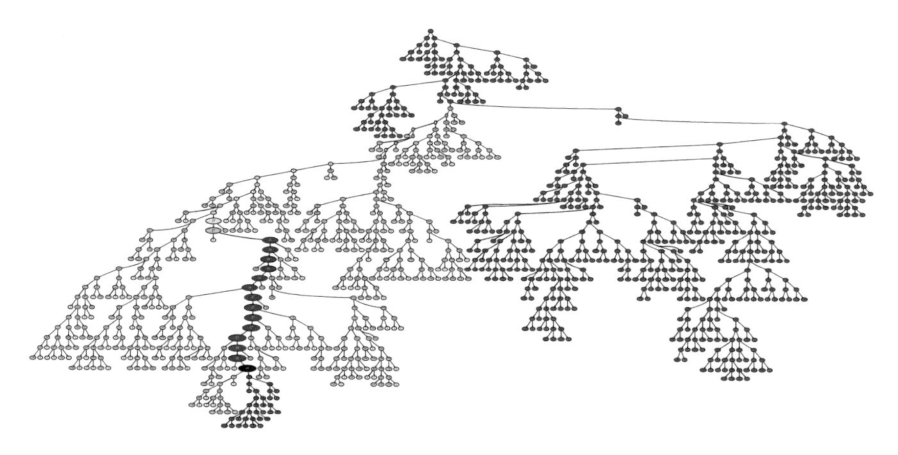

Fig. 8.7 Example of incremental evaluation [15]. Parent tree is modified by crossover replacing code with inserted subtree (red). Replaced and new code are both evaluated on the test set (48 tests). As they are different, the next node above the crossover point is evaluated, taking the 48 values returned by the original and new code (together with its other argument from the unchanged code). Here too evaluation in the parent and (putative) child are different, so evaluation proceeds up the tree towards its root node (see also Fig. 8.6). The chain of evaluated nodes is in colour [19]. The size and numbers in each node gives the number of test cases where the evaluation of the parent and (putative) child are not identical. Their average evaluation difference is indicated on a log scale by the node's colour. Average differences greater than 0.01 are shown with dark colours, less than 0.01 by brighter colours. Brightest yellow shows smallest non-zero difference (RMS 3.1 10^{-10}). If, as here, parent and child evaluations are identical before reaching the root node, the remainder of the evaluation is not needed (gray nodes) and is skipped and instead fitness is copied from the parent

And therefore the child's fitness must be equal to that of its mum. Note we still have not gone near the child and indeed we have finished with the dad.

If the two vectors are not identical, we proceed up the mum tree evaluating side subtrees and nodes on the path to the root until either we reach a point where the values in the mum and the values the child would have been identical or we reach the root. If we reach the root, the child's fitness is calculated from the values in its vector of evaluations for each test case. Again we do not need to create the child to do this.

In very big trees, populations are often highly converged and children often inherit the same fitness values as their parents. In which case, fitness first evaluation can give orders of magnitude savings in evaluation time.

Table 8.1 gives details of our GP.

8.3 Avoiding Effort Wasted on Poor Fitness Individuals

Whereas the previous approaches, described in Sects. 8.2.1 and 8.2.2, speed up genetic programming by use of more powerful hardware, we have implemented

Table 8.1 Evolution of Sextic polynomial [7] symbolic regression binary trees using GPquick's one byte per opcode

Terminal set	X, 250 constants between −0.995 and 0.997
Function set	MUL ADD DIV SUB
Fitness cases	48 fixed input −0.97789–0.979541 (randomly selected from −1.0 to +1.0). For simplicity, we use all the same test cases in each generation, although of course, testing can be reduced [5, 21] or made dynamic [18] Target $y = xx(x-1)(x-1)(x+1)(x+1)$
Selection	Tournament size 7 with fitness = $\frac{1}{48}\sum_{i=1}^{48} \lvert GP(x_i) - y_i \rvert$
Population	500 binary trees. Panmictic (fully mixed), non-elitist, distinct (non-overlapping) generations.
Parameters	Initial population ramped half and half [7], depth between 2 and 6. 100% unbiased subtree crossover. 70 000 generations

a fitness first scheme which speeds up GP by 14% by doing less work. (Fitness first could be widely applicable in evolutionary computing, however only when constructing members of the population is expensive compared to fitness evaluation is it likely to be useful.) For simplicity our implementation ensures that it produces identical results. That is, given the same pseudo random number seed, the population at each generation in the new implementation is identical to that given before.

Early in GP runs at each generation many poor individuals are created (see Fig. 8.8). All Evolutionary Algorithm (EA) selection schemes aim to ensure poor individuals are less likely to be selected to have children themselves. (See example with a population of five in Fig. 8.5.) Since childless individuals have no impact on the future course of the run, it is wasteful to create such individuals.

Apart from Baker's Stochastic Uniform Selection (SUS) [1], commonly used selection schemes, such as tournament selection, allocate children independently. Thus, even later in the run, when many programs have the same fitness, there will be some parents who by chance get more than the average number of children and some who get less. With two parent crossover, on average each member of the current population gets two children. In the limit of large converged populations (containing M individuals) on average there will be $e^{-2}M$ individuals which are never selected to have children (see right hand side of Fig. 8.8). If we consider just the first parent in crossover, or 100% one parent mutation, then this rises to $e^{-1}M$.

As Fig. 8.8 shows, delaying crossover until after fitness selection can save creating more than half the population during the early part of a run. Even later, when convergence ensures almost the whole population has the same fitness, 14% (e^{-2}) of the population need not be created. With very large trees, run time can be dominated by crossover (see Fig. 8.3), thus run time savings are possible by avoiding complete generation of poor fitness individuals.

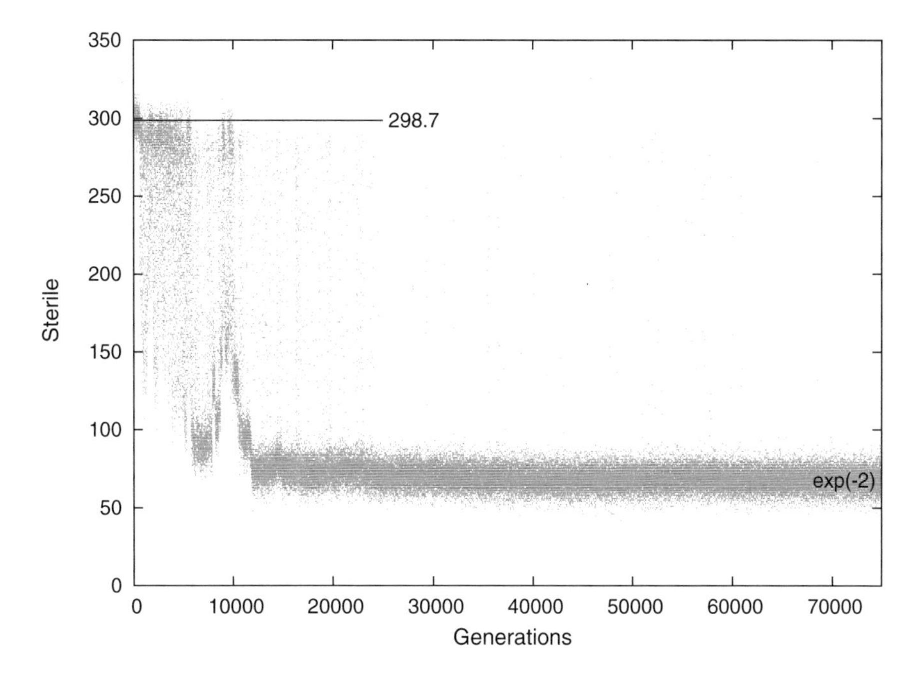

Fig. 8.8 Evolution of number in population without children in next generation. 100% two parent crossover, 7-tournament, pop = 500

8.4 Asymmetry of GP Subtree Crossover

We use Koza's two point subtree crossover [7] but for simplicity with both crossover points chosen uniformly at random. That is, we do not include a bias in favour of internal nodes.

Figure 8.9 shows the dramatic imbalance in the contributions of the two programs chosen to be parents for the new individuals (note log scale). For example, in generation 15 000 the root donating trees (mums) supply more than a thousand times as many opcodes as the dads.

The lower (red) solid line in Fig. 8.9 plots the running mean smoothed over 100 generations of the number of inserted opcodes from each dad program. After generation 15 000 it changes little, and averages 275.4 opcodes. However the distribution of inserted subtree sizes varies widely in each generation and between generations (blue dots). It has a long tail with the mean being typically more than three times the median. The dad long tailed distribution has some impact on run time, with some trees taking far longer to evaluate for fitness than others, making it harder to distribute work evenly between threads on multi-core CPUs. (Section 8.8.1 considered how often cores are not being used.) In contrast the number of opcodes inherited from mum (top line in Fig. 8.9) closely follows the total tree size and even after generation 15 000 continues to bloat.

8.4.1 Last Child Inplace Dad-Less Crossover

Initially the populations are very variable and, with strong selection, breeding is concentrated in a few fit parents. As the populations starts to converge, there are more parents (with fewer children each). In each generation, as each child is created, eventually for each parent, there is only one child left to be created. (Locks are used to ensure multi-threaded code neither skips anyone nor creates any child twice). On reaching the last child for a root donating parent, instead of copying the code into the child (see Fig. 8.10), the buffer holding the parent's genome is unhooked from the parent and passed to the child. This saves copying the first part of the child (see Fig. 8.11).

As we saw in Fig. 8.9, the second parent (dad) donates only a tiny fraction of the opcodes in the child. Therefore we extract and save all the subtrees which will be inserted later. This is relatively cheap and is done (in the sequential code) before the

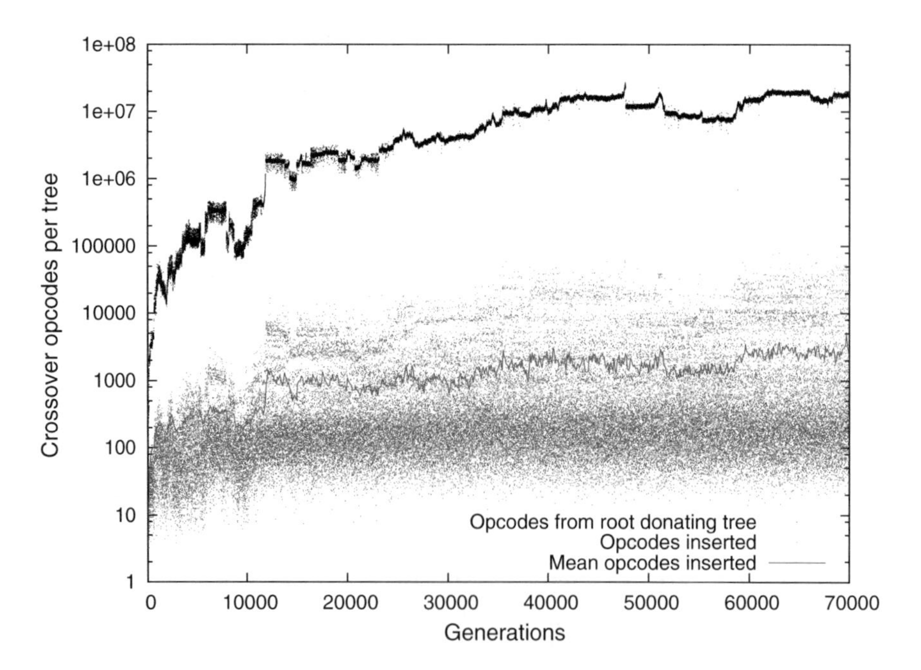

Fig. 8.9 Evolution of number opcodes from each parent. Mums top line. Dads blue lower cloud. Note log vertical scale

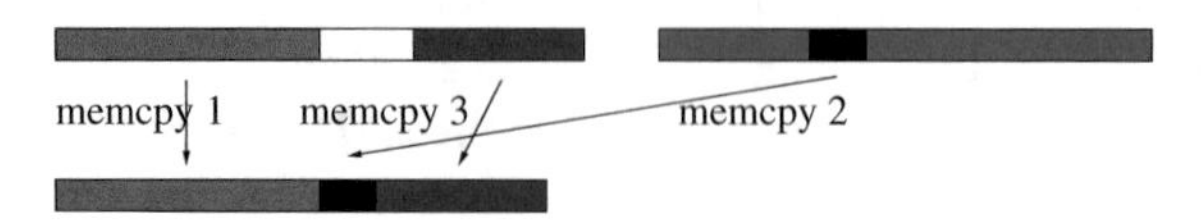

Fig. 8.10 Andy Singleton's GPquick [31] subtree crossover requires three memcpy buffer copies: (1) root segment of donating parent (mum, red/brown) is copied to offspring buffer. (2) subtree from second parent (dad, blue/black) is copied to offspring. (3) tail (brown) of 1st parent copied to child

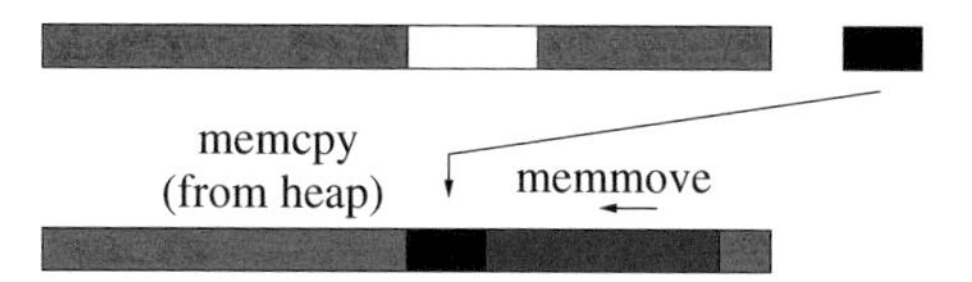

Fig. 8.11 Inplace subtree crossover. Offspring is last child of 1st parent and reuses its buffer. Only subtree to be inserted (black) of 2nd parent (dad) is kept. (1) Dad subtree overwrites mum's buffer. (2) In 71% of children the subtree to be remove (white) and to be inserted (black) are different sizes, and so memmove is used to shuffle the second part of mum's buffer (brown) up or down

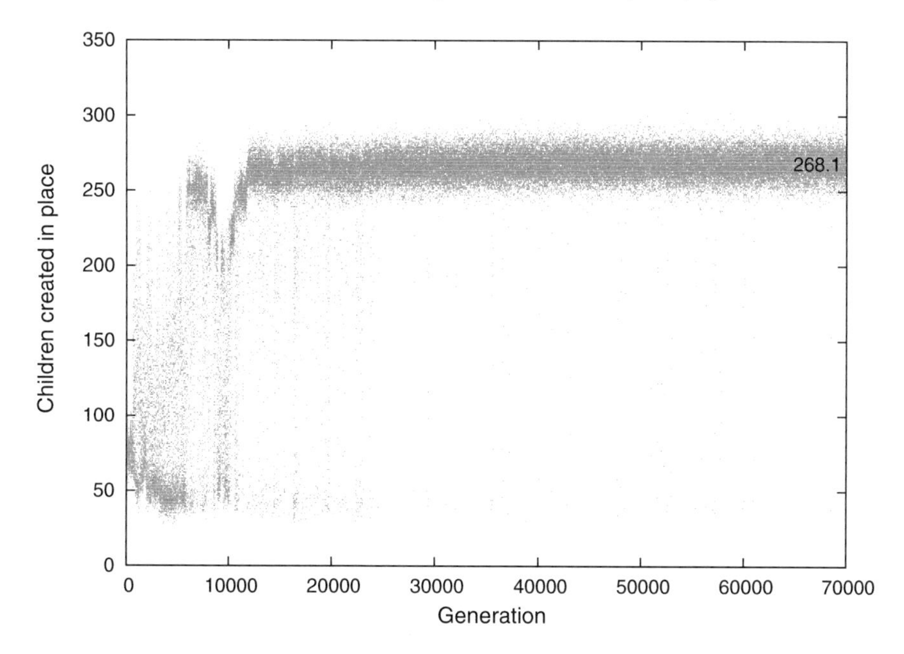

Fig. 8.12 Number of times per generation when creating non-sterile children in the next, the root donating parent (mum) has only one more child to create and so crossover can reuse part of its genome. Pop = 500. See Sect. 8.4.1 and Fig. 8.11

bulk of the crossover operations are done using the root donating parents (mums) in multi-threaded code. This simple step allows the mum's last child crossover short cut (Fig. 8.11) to be used about twice as often.

Notice whilst fitness convergence reduces the number of childless members of the population (Sect. 8.3), here it helps: as spreading the breeding effort, means there are more parents in general, and thus more cases where a mum has only one child left to be created. That is, convergence increases the number of times inplace crossover optimisation can be applied. Figure 8.12 shows later in the run as the population converges and there are more parents with children, the number of inplace crossovers rises, so that on average 268.1 ($\lesssim M(1-e^{-2})(1-e^{-1})$) crossovers are done inplace per generation.

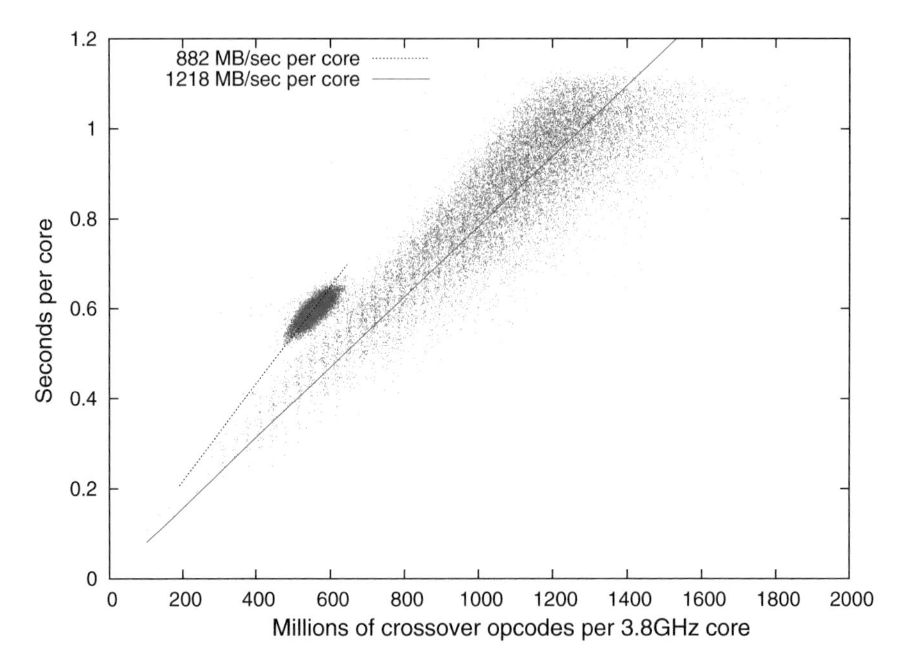

Fig. 8.13 Time per thread to create children using fatherless (left, red) and traditional (right, blue) crossovers v. The number of opcodes the thread processes (see Sect. 8.4.1). To reduce clutter just generations 69 000–70 000 are plotted. 16 core 3.8 GHz desktop

In about one third (28.9%) of cases, the removed subtree and inserted subtree are the same size. If so, the mum's buffer can be simply over written with the inserted code (from the dad). However most (71.1%) of the time they are not the same size and the buffer must be shuffled either up or down to take account of the difference in the subtree sizes (see Fig. 8.11). This shuffling is done using memmove, rather than memcpy. (See also Sect. 8.5). Figure 8.12 confirms, by excluding the dads from crossover, we can use the inplace short cut more than half the time.

The large blue cloud in Fig. 8.13 shows the time originally taken by each of 16 threads to perform crossover of the whole of the current generation late in the run. The tight red cluster of dots show the same populations after crossover has been optimised to: (1) ignore individuals which will not have children (saving about 13.5%) and (2) where possible, modifying chromosomes inplace. Figure 8.13 confirms we are reducing the volume of opcodes copied by crossover by almost a half (48.1%). This leads to a reduction in the total time taken by the crossover threads by about a quarter (24.4%).

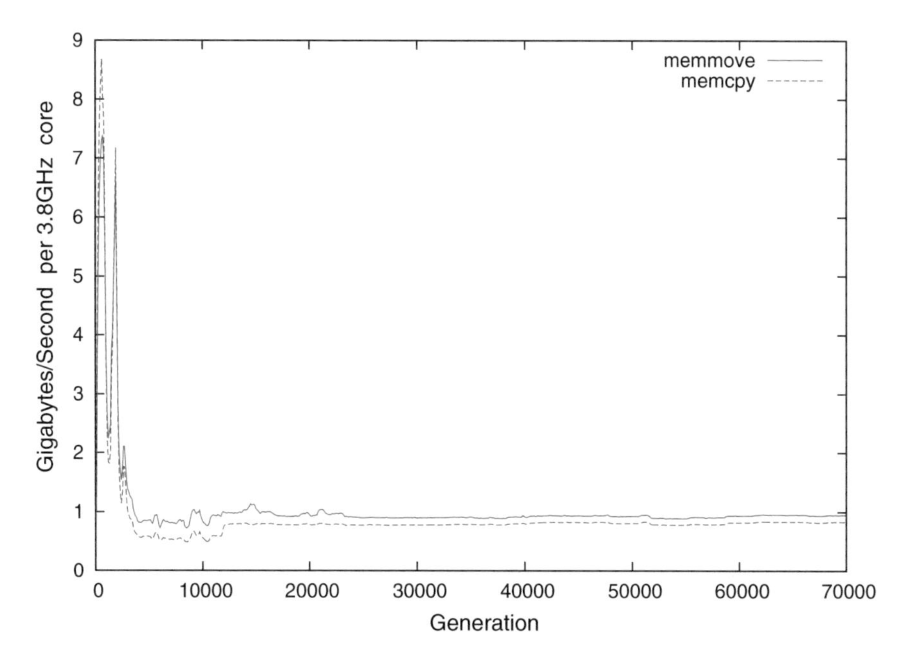

Fig. 8.14 Evolution of speed of memmove and memcpy as used in GPquick crossover. It appears the initial high speed of both is due to GP trees not exceeding the cache size, 16.5 MB. Plots are smoothed running means of 100 generations. Note traditionally bandwidth counts each byte moved or copied twice, i.e. a byte into the CPU and a byte out to memory

8.5 Efficiency of Memmove V. Memcpy

Although much has been made of the efficiency of memcpy compared to that of memmove, with the GCC 9.3.1 g++ compiler and version 2.17 of the GNU C run time library, for our new crossover implementation we found little difference (see Fig. 8.14). Indeed instrumenting the memmove operation and the corresponding memcpy, shows memmove to be 14% faster. On average at the end of the run memmove moves 970MB/second per core while memcpy copies 851MB/sec per core (on a 3.80 GHz Intel i7-9800X desktop). Note that these are in place measurements, rather than standalone benchmarks and so memmove has on average slightly more work.

8.6 Speed of Fitness First and Incremental Fitness

As described in Sect. 8.2.3, our incremental fitness evaluation [15], which evaluates side-effect free trees from the crossover point towards the root, can be readily adapted to evaluate the child via its parents. Apart from adapting pointers to the crossover

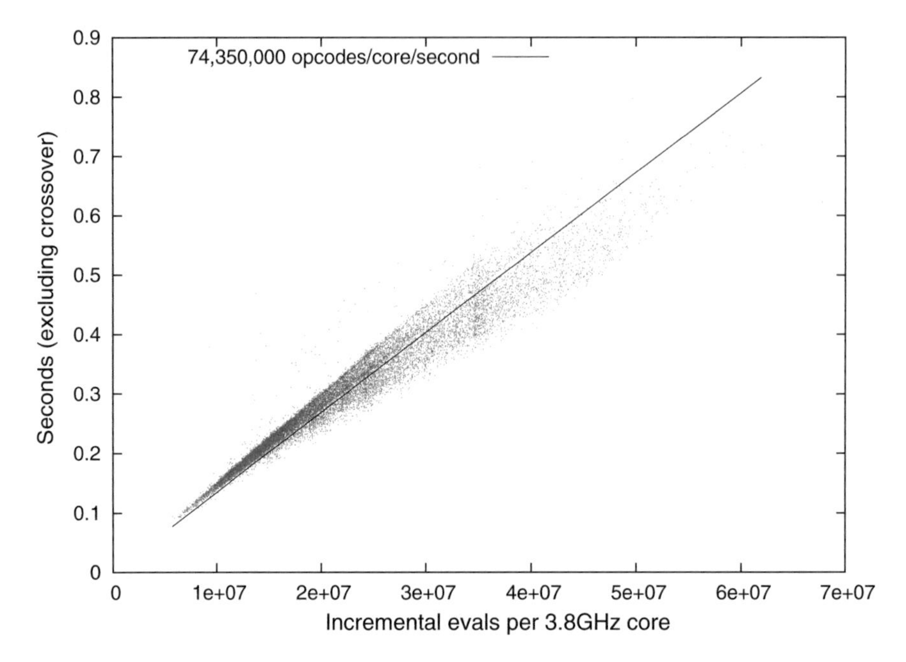

Fig. 8.15 Time taken each generation by each thread to calculate fitness against the number of opcodes the thread processes. Note incremental fitness evaluation using the child's two parents before the child is created. Scatter plot, 16 threads, generations 69 000–70 000

points in the parents, rather than in their child, little is changed. As expected, Fig. 8.15 shows the time taken to find the fitness of the whole of the current generation depends linearly on the number of opcodes that have to be evaluated. Note inparticular moving from incremental evaluation of the children to evaluating them by using only their parents has made little difference, see lower dash and dotted traces in Fig. 8.3. (The fitness results are of course identical.)

8.7 Mathematical Model of Number of Parents

Section 8.3) has already shown a model of crossover which predicts the number of members a population with near uniform fitness which do not have children in the next generation will be $e^{-2}M$. Figure 8.8, confirms the model essentially holds after generation 15 000 even though there remain a few members of the population with an atypical fitness value. (See also Fig. 8.1.)

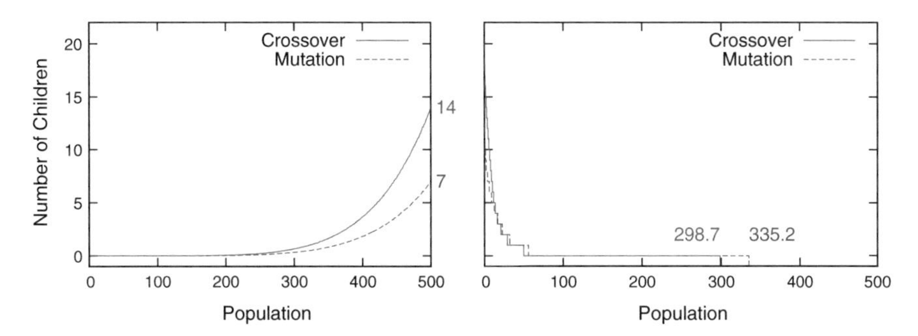

Fig. 8.16 Left: Offspring v. rank. Expected number of children with tournament size T=7 in initial diverse populations [2, 20]. Pop=500. Single parent mutation (dashed line) not used but shown for comparison. Right: same data as histograms. E.g. on average 298.7 members of the population (with crossover) have no children, 49.9 have one child, 29.1 two and so on

8.7.1 Number of Parents Initially and in Diverse Populations

Where there is a fitness gradient across the population, a wide variety of selection schemes will allocate children to the best members of the population. This means even with two parent operations, like crossover, there will be many low fitness or just unlucky members of the population, whose genetic material will be lost.

Goldberg's selection pressure [4] of commonly used fitness selection schemes has been mathematically analysed by Blickle [2], and ourselves [20, p. 185] giving, in a diverse population, the chance of the rth best individual in the population winning the next tournament as $(r/M)^T - ((r-1)/M)^T$ (see Fig. 8.16). Assuming distinct non-elitist generations and T-tournament selection, on average the best member of the population will be selected to be a first parent T times. Using crossover there are two parents, so parents have twice as many children. Thus, the best in the population has on average $2T = 14$ children (see left of Fig. 8.16). Even in modest population sizes, the worst member of the population is unlikely to have children.

A Monte Carlo simulation predicts almost 60% of random populations with a tournament size of seven will not have children, see Fig. 8.16. This is good agreement with many populations up to about generation 5000, i.e. before they near fitness convergence (see 298.7 of 500, in Figs. 8.8 and 8.16).

8.8 Multi-threading Implementation Issues

To minimise memory consumption, we process children whose parents have only one child left be delt with before the others [12]. This avoids having to store both the current and the next population at the same time. As children are created, their parents are moved between two queues. One queue is for parents with one child left to process and another queue is for parents with two or more children yet to be created. When a parent's last child has been created, the parent can be deleted and the memory it occupied can be freed and thus be used by new children in the next generation. As we reported earlier [12], with the usual crossover and fitness evaluation order, M+2N memory buffers are needed. Where M is the population size and there are N threads. The factor of two comes from using two parent crossover. (If using only single parent mutation, M+N memory buffers would be needed.)

By using fatherless crossover, M+2N, can be reduced to M+1N. Although fatherless crossover, Sects. 8.4 and 8.4.1, does require storing the subtrees to be inserted on the heap. However typically the opcodes inherited from the dads occupy less than a megabyte (see Fig. 8.9).

The two multi-threaded queues [12] give an easy way of recognising mums with only one child left to create and so help implementing inplace crossover, see Sect. 8.4.1) and Figs. 8.10, 8.11). Also, as inplace crossover automatically shares the memory used by the parent and the offspring, in practice memory consumption is reduced to approximately $(1 - e^{-1})$M+N = 0.63M+N. That is, although we still have to allow for N threads operating simultaneously: population fitness convergence, not creating low fitness individuals who will not have children, fatherless crossover and inplace crossover, together (as well as speeding up GP) reduce memory consumption by about a third.

Although we know on which of the two queues parents must be placed [12], we are still free to decide where in the given queue they are to be. As yet we have not exploited this ordering freedom. In future there may be modest saving to be made by better scheduling work between the available threads. (We return to this in Sect. 8.8.2.)

8.8.1 Idle Threads

Figures 8.17 and 8.18 show the total thread idle time on a 16 core desktop. Figure 8.18 shows the average waiting time as a fraction of the elapse time for each set of 16 threads in that generation. To improve visibility, the plots have been smoothed by taking running averages over 100 generations.

In the original scheme (blue dashed lines) multiple threads performed crossover and evaluated fitness [15]. I.e. children were created and their fitness was immediately calculated, as an indivisible unit, by the same thread. (Note crossover was performed to create 100% of each population.) In the new scheme, crossover of only the part of the *next* generation which has children is done (red lines with crosses). Fitness evaluation is unchanged. Since crossover and fitness now operate on different individuals, they are separated, and each is done by their own set of threads. For simplicity the two sets do not overlap. I.e. the fitness threads synchronise together and then the crossover threads synchronise together. In principle the two types of threads could be intermingled, but this would complicate the implementation.

Thus, in the original scheme, there is only one synchronisation point at the end of each generation, where idle threads are forced to wait. Whereas there are two synchronisation points in the new scheme. (Hence the three sets of lines in Figs. 8.17 and 8.18.)

In both schemes, the later stages of the run are dominated by the crossover time (see top two lines in Fig. 8.3). However crossover time is much more predictable and uniform than the time to do fitness evaluation (where the longest fitness evaluation can exceed the average by a factor of 100 or more). Fitness evaluation is simply scheduled by the next free thread taking the next individual. Whereas the order of the crossover threads is dictated by Koza's algorithm to minimise buffer usage [8, pp. 1044–1045], [12, 13] (see previous section).

The more uniform duration of the crossover tasks means thread idle time, as a fraction of total time (Fig. 8.18), is low. The wide variation in fitness evaluation time leads to proportionately more wasted thread idle time. However this is mitigated in bloated runs by the great speed of incremental fitness evaluation compared to the time taken to create enormous trees. For example, on average over the last 100 generations, GP was unable to use 39%, of the 16 core computer during fitness evaluation (top trace in Fig. 8.18), whilst for the new crossover it was 1% unused.

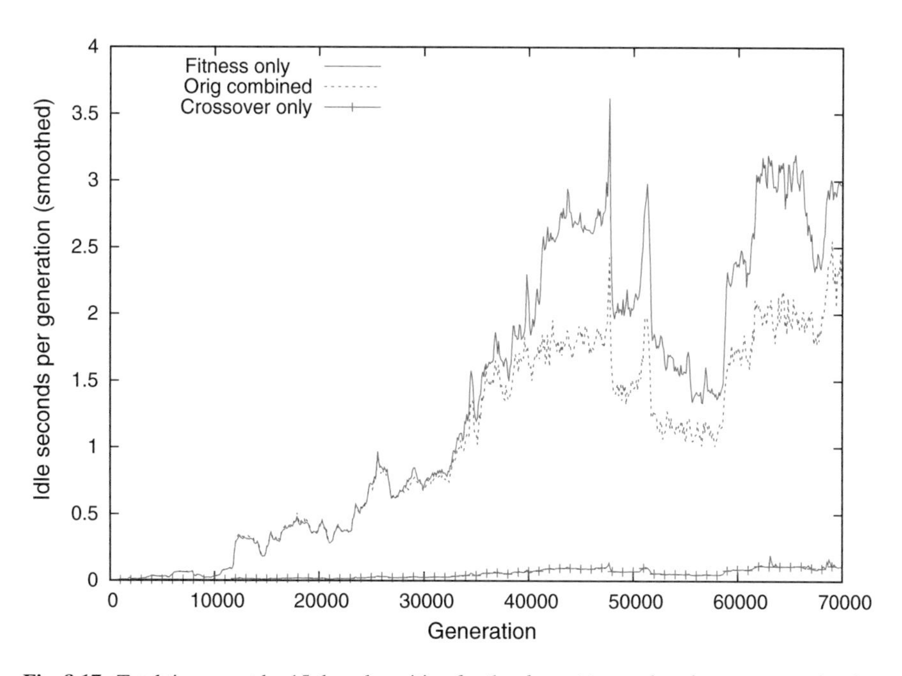

Fig. 8.17 Total time spent by 15 threads waiting for the slowest to synchronise per generation (on 16 core 3.8 GHz desktop). In the original implementation (dashed blue line) the original crossover and our incremental [15] fitness evaluation were performed together. In the new crossover and fitness are separated, leading to two synchronisation steps per generation and two sets of idle threads (solid red lines). See also Fig. 8.18

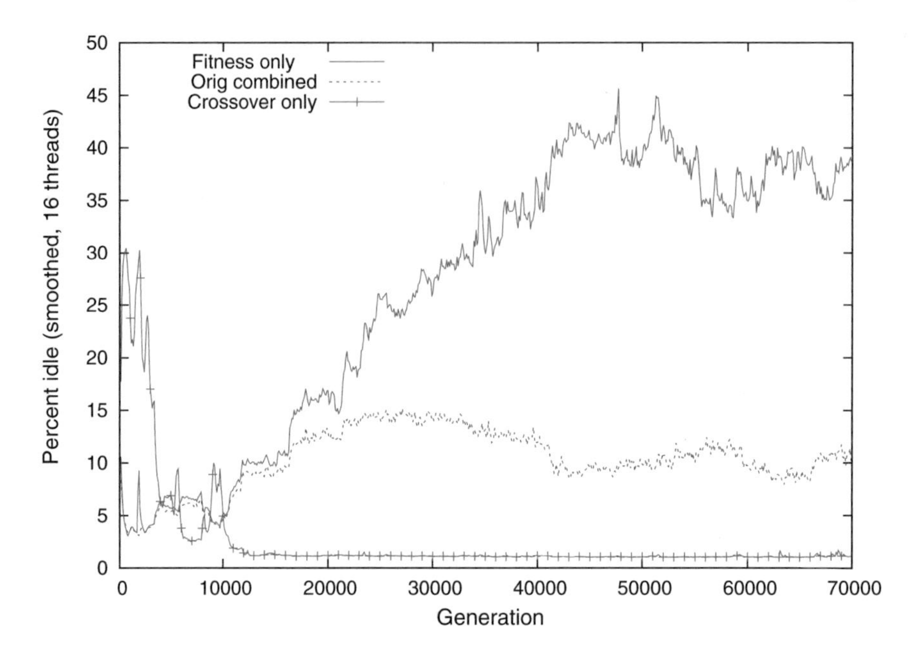

Fig. 8.18 Time spent by 15 threads waiting for the slowest as a fraction of time taken by all 16. Data as Fig. 8.17 but expressed as percentages

8.8.2 Future Work: Predicting Thread Execution Time

As mentioned in the previous section, when a thread finishes a task it takes the next free task and begins processing it. Idle time comes from threads running out of tasks at different times. When tasks take different lengths of time, there may be practical savings from more proactive scheduling. Since the threads are (assumed to be) homogeneous, a simple heuristic of starting with the longest tasks (spread across all the threads) and then moving to progressively shorter tasks, may be sufficient. E.g. sort the tasks into execution time order and then run as now.

Crossover time can be readily predicted from the amount of memory to be moved (memmove) or copied (memcpy). Given the size of individuals and the location of crossover points, both can be calculated in advance. So, for simplicity treating memmove and memcpy as the same, to minimise idle time, we might want to order the crossover queues to put the largest children first. However, to maximise runtime savings from inplace crossover, we might want to try to schedule crossovers so that children with the largest root segments are the last to be done for their mums. Alternatively to save memory, we might want to do them as soon as possible. (In [12] we treated all trees as being the same size.)

Fitness evaluation time is very variable and hard to predict, as, even though it is proportional to the number of opcodes to be evaluated (Fig. 8.15), the number of evals is only known after the evaluation. It may be possible even a crude model might help. E.g. guess that a large (or very different) subtree to be inserted, will cause more disruption and hence require more evals, than a smaller or more similar one. Fitness first execution times can be very variable and, with 16 threads, a single evaluation can take as long as the rest of the population (spread over 15 threads). Given this, there may be only marginal gain from clever scheduling. As the variation gets still bigger it might be, for very time consuming individuals, worthwhile to spread their fitness evaluation across multiple threads.

8.9 Conclusions

Although we have couched our work in GP terms, the memory savings hold for evolutionary algorithms with crossover or with mutation alone. Where EA chromosomes are enormous and (changes in) fitness can be quickly calculated, these ideas of reversing the order of fitness calculation and offspring creation, might also be beneficial.

For a typical small GP population (500 trees) on a 16 node desktop, memory use can be reduced by about a third. On that desktop we have performance *equivalent* to 692 Giga GPop/s ($6.92\ 10^{11}$ GP operations per second) which is more than four times the performance that we claimed as a record [16] for a single computer GP system and that was a 48 core cluster server.

We have shown it is practical to delay subtree crossover until after fitness evaluation and so only create GP trees which themselves will carry genetic material into subsequent generations. Typically early in GP runs, tournament selection gives a very high selection pressure, meaning there are many trees of low fitness which do not have children. In any evolutionary algorithm, by reversing the usual order of program evaluation and creation, it is no longer necessary to create low fitness individuals. This can save a large fraction of the memory to store them. Even later in GP runs, when fitness convergence may spread children more evenly, and the cost of creating new GP trees may exceed the cost of fitness evaluation, the saving can be worthwhile.

Even when trees are large, the asymmetry of GP subtree crossover means, the code to be inserted into the next generation, (i.e. all the subtrees from each father) is small. Indeed it may fit into fast cache memory. These subtrees can be extracted from the population before the bulk crossover operations. This simplifies the rest of the crossover operations, as they now only use one parent from the population (i.e. they are fatherless). This can be beneficial in terms of freeing memory early and reducing crossover effort.

The new GPQuick code is available in http://www.cs.ucl.ac.uk/staff/W.Langdon/ftp/gp-code/GPinc.tar.gz

Acknowledgements I would like to thank Stephan Winkler, Sara Silva, Bill Tozier, other people at GPTP and anonymous reviewers. This work was inspired by conversations at Dagstuhl Seminar 18052 on Genetic Improvement of Software [26]. Funded by EPSRC grant EP/P005888/1.

References

1. Baker, J.E.: Reducing bias and inefficiency in the selection algorithm. In: Grefenstette, J.J. (ed.) Proceedings of the Second International Conference on Genetic Algorithms and their Application, pp. 14–21. Lawrence Erlbaum Associates, Cambridge, MA, USA (1987)
2. Blickle, T.: Theory of evolutionary algorithms and application to system synthesis. Ph.D. thesis, Swiss Federal Institute of Technology, Zurich, Switzerland (1996). http://dx.doi.org/10.3929/ethz-a-001710359
3. de Melo, V.V., Fazenda, A.L., Sotto, L.F.D.P., Iacca, G.: A MIMD interpreter for genetic programming. In: Castillo, P.A., Jimenez Laredo, J.L., Fernandez de Vega, F. (eds.) 23rd International Conference, EvoApplications 2020, LNCS, vol. 12104, pp. 645–658. Springer, Seville, Spain (2020). URL http://dx.doi.org/10.1007/978-3-030-43722-0_41
4. Goldberg, D.E.: Genetic Algorithms in Search Optimization and Machine Learning. Addison-Wesley (1989)
5. Guizzo, G., Petke, J., Sarro, F., Harman, M.: Enhancing genetic improvement of software with regression test selection. In: van Deursen, A., Xie, T., Dieste, N.J.O. (eds.) Proceedings of the International Conference on Software Engineering, ICSE 2021. IEEE (2021). http://dx.doi.org/10.1109/ICSE43902.2021.00120. Winner ACM SIGSOFT Distinguished Artifact Award
6. Hrbacek, R., Sekanina, L.: Towards highly optimized cartesian genetic programming: from sequential via SIMD and thread to massive parallel implementation. In: C. Igel, D.V. Arnold, C. Gagne, E. Popovici, A. Auger, J. Bacardit, D. Brockhoff, S. Cagnoni, K. Deb, B. Doerr, J. Foster, T. Glasmachers, E. Hart, M.I. Heywood, H. Iba, C. Jacob, T. Jansen, Y. Jin, M. Kessentini, J.D. Knowles, W.B. Langdon, P. Larranaga, S. Luke, G. Luque, J.A.W. McCall, M.A. Montes de Oca, A. Motsinger-Reif, Y.S. Ong, M. Palmer, K.E. Parsopoulos, G. Raidl, S. Risi, G. Ruhe, T. Schaul, T. Schmickl, B. Sendhoff, K.O. Stanley, T. Stuetzle, D. Thierens, J. Togelius, C. Witt, C. Zarges (eds.) GECCO '14: Proceedings of the 2014 conference on Genetic and evolutionary computation, pp. 1015–1022. ACM, Vancouver, BC, Canada (2014). URL http://dx.doi.org/10.1145/2576768.2598343
7. Koza, J.R.: Genetic Programming: On the Programming of Computers by Means of Natural Selection. MIT Press, Cambridge, MA, USA (1992). http://mitpress.mit.edu/books/genetic-programming
8. Koza, J.R., Andre, D., Bennett III, F.H., Keane, M.: Genetic Programming III: Darwinian Invention and Problem Solving. Morgan Kaufmann (1999). http://www.genetic-programming.org/gpbook3toc.html
9. Langdon, W.B.: Long-term evolution of genetic programming populations. In: Proceedings of the Genetic and Evolutionary Computation Conference Companion, GECCO '17, pp. 235–236. ACM, Berlin (2017). http://dx.doi.org/10.1145/3067695.3075965
10. Langdon, W.B.: Parallel GPQUICK. In: C. Doerr (ed.) GECCO '19: Proceedings of the Genetic and Evolutionary Computation Conference Companion, pp. 63–64. ACM, Prague, Czech Republic (2019). http://dx.doi.org/10.1145/3319619.3326770
11. Langdon, W.B.: Genetic improvement of genetic programming. In: Brownlee, A.S., Haraldsson, S.O., Petke, J., Woodward, J.R. (eds.) GI @ CEC 2020 Special Session, p. paper id24061. IEEE Computational Intelligence Society, IEEE Press, internet (2020). http://dx.doi.org/10.1109/CEC48606.2020.9185771
12. Langdon, W.B.: Multi-threaded memory efficient crossover in C++ for generational genetic programming. SIGEVOLution newsletter of the ACM Special Interest Group on Genetic and Evolutionary Computation **13**(3), 2–4 (2020). URL http://dx.doi.org/10.1145/3430913.3430914
13. Langdon, W.B.: Multi-threaded memory efficient crossover in C++ for generational genetic programming (2020). http://arxiv.org/abs/2009.10460
14. Langdon, W.B.: Fitness first and fatherless crossover. In: Proceedings of the Genetic and Evolutionary Computation Conference Companion, GECCO '21. ACM, Internet, pp. 253–254 (2021). http://dx.doi.org/10.1145/3449726.3459437.

15. Langdon, W.B.: Incremental evaluation in genetic programming. In: Hu, T., Lourenco, N., Medvet, E. (eds.) EuroGP 2021: Proceedings of the 24th European Conference on Genetic Programming, LNCS, vol. 12691, pp. 229–246. Springer, Virtual Event (2021). http://dx.doi.org/10.1007/978-3-030-72812-0_15
16. Langdon, W.B., Banzhaf, W.: Continuous long-term evolution of genetic programming. In: Fuechslin, R. (ed.) Conference on Artificial Life (ALIFE 2019), pp. 388–395. MIT Press, Newcastle (2019). http://dx.doi.org/10.1162/isal_a_00191
17. Langdon, W.B., Banzhaf, W.: Faster genetic programming GPquick via multicore and advanced vector extensions. Technical Report RN/19/01, University College, London, London, UK (2019). http://www.cs.ucl.ac.uk/fileadmin/user_upload/avx_rn1901.pdf
18. Langdon, W.B., Harman, M.: Optimising existing software with genetic programming. IEEE Trans. Evolut. Comput. **19**(1), 118–135 (2015). http://dx.doi.org/10.1109/TEVC.2013.2281544
19. Langdon, W.B., Petke, J., Clark, D.: Dissipative polynomials. In: Veerapen, N., Malan, K., Liefooghe, A., Verel, S., Ochoa, G. (eds.) 5th Workshop on Landscape-Aware Heuristic Search, GECCO 2021 Companion. ACM, Internet, pp. 1683–1691 (2021). http://dx.doi.org/10.1145/3449726.3463147
20. Langdon, W.B., Poli, R.: Foundations of Genetic Programming. Springer (2002). http://dx.doi.org/10.1007/978-3-662-04726-2
21. Lim, M., Guizzo, G., Petke, J.: Impact of test suite coverage on overfitting in genetic improvement of software. In: Galeotti, J.P., Sharif, B. (eds.) 12th International Symposium on Search Based Software Engineering SSBSE 2020, LNCS, vol. 12420, pp. 188–203. Springer, Bari, Italy (2020). http://dx.doi.org/10.1007/978-3-030-59762-7_14
22. Petke, J.: Constraints: The future of combinatorial interaction testing. In: 2015 IEEE/ACM 8th International Workshop on Search-Based Software Testing, pp. 17–18. Florence (2015). http://dx.doi.org/10.1109/SBST.2015.11
23. Petke, J., Haraldsson, S.O., Harman, M., Langdon, W.B., White, D.R., Woodward, J.R.: Genetic improvement of software: a comprehensive survey. IEEE Trans. Evolut. Comput. **22**(3), 415–432 (2018). http://dx.doi.org/10.1109/TEVC.2017.2693219
24. Petke, J., Harman, M., Langdon, W.B., Weimer, W.: Using genetic improvement and code transplants to specialise a C++ program to a problem class. In: Nicolau, M., Krawiec, K., Heywood, M.I., Castelli, M., Garcia-Sanchez, P., Merelo, J.J., Rivas Santos, V.M., Sim, K. (eds.) 17th European Conference on Genetic Programming, LNCS, vol. 8599, pp. 137–149. Springer, Granada, Spain (2014). http://dx.doi.org/10.1007/978-3-662-44303-3_12
25. Petke, J., Harman, M., Langdon, W.B., Weimer, W.: Specialising software for different downstream applications using genetic improvement and code transplantation. IEEE Trans. Softw. Eng. **44**(6), 574–594 (2018). http://dx.doi.org/10.1109/TSE.2017.2702606
26. Petke, J., Le Goues, C., Forrest, S., Langdon, W.B.: Genetic improvement of software: Report from dagstuhl seminar 18052. Dagstuhl Rep. **8**(1), 158–182 (2018). http://dx.doi.org/10.4230/DagRep.8.1.158
27. Poli, R.: TinyGP. TinyGP GECCO 2004 competition (2004). http://www.cs.ucl.ac.uk/staff/W.Langdon/ftp/papers/poli04__tinyg.pdf
28. Poli, R., Langdon, W.B.: Sub-machine-code genetic programming. In: Spector, L., Langdon, W.B., O'Reilly, U.M., Angeline, P.J. (eds.) Advances in Genetic Programming 3, chap. 13, pp. 301–323. MIT Press, Cambridge, MA, USA (1999). http://www.cs.ucl.ac.uk/staff/W.Langdon/aigp3/ch13.pdf
29. Poli, R., Langdon, W.B.: Running genetic programming backward. In: Yu, T., Riolo, R.L., Worzel, B. (eds.) Genetic Programming Theory and Practice III, Genetic Programming, vol. 9, Chap. 9, pp. 125–140. Springer, Ann Arbor (2005). http://dx.doi.org/10.1007/0-387-28111-8_9
30. Poli, R., Langdon, W.B., McPhee, N.F.: A field guide to genetic programming. Published via http://lulu.com http://www.gp-field-guide.org.uk (2008). http://www.gp-field-guide.org.uk. (With contributions by J. R. Koza)

31. Singleton, A.: Genetic programming with C++. BYTE pp. 171–176 (1994). http://www.assembla.com/wiki/show/andysgp/GPQuick_Article
32. White, D.R., Arcuri, A., Clark, J.A.: Evolutionary improvement of programs. IEEE Trans. Evolut. Comput. **15**(4), 515–538 (2011). http://dx.doi.org/10.1109/TEVC.2010.2083669

Chapter 9
Designing Multiple ANNs with Evolutionary Development: Activity Dependence

Julian Francis Miller

Abstract We use Cartesian genetic programming to evolve developmental programs that construct neural networks. One program represents the neuron soma and the other the dendrite. We show that the evolved programs can build a network from which *multiple* conventional ANNs can be extracted each of which can solve a different computational problem. We particularly investigate the utility of activity dependence (AD), where the signals passing through dendrites and neurons affect their properties.

9.1 Introduction

Although ANNs were originally inspired by the brain [16], most do not use evolution and especially not development. A major weakness of ANN models is the "synaptic dogma" in which learned knowledge is held solely in connection strengths (i.e. weights). This gives rise to the fundamental problem of "catastrophic forgetting" [8, 15, 26]. This occurs when an ANN loses it ability to solve a earlier problem when it is re-trained on a new one. This is to be expected when the learned information is only encoded in the weights as it is precisely these that are changed when the network is trained. Another problem is that memory in brains is not even directly related to synaptic strengths. This is because synapses are not fixed but are constantly pruned away and replaced by new synapses during learning [29]. Also a large body of research indicates that learning and environmental interaction are strongly related to *structural* changes in neurons. Animals reared in complex environments where active learning is taking place, have an increased density of dendrites and synapses [11]. Breeding songbirds undergo an increase in the number, size and spacing of neurons [32]. Furthermore, a study of London taxi drivers, showed their hippocampi were significantly larger relative to those of control subjects [14].

Since the emergence of deep learning there has been renewed interest in artificial neural network (ANN) approaches to artificial intelligence (AI). There are two main approaches. The manual approach and an AI generating algorithm (AI-GA). The

J. F. Miller (✉)
University of York, Heslington, York YO10 5DD, UK

W. Banzhaf et al. (eds.), *Genetic Programming Theory and Practice XVIII*,
Genetic and Evolutionary Computation,
https://doi.org/10.1007/978-981-16-8113-4_9

former is adopted by the vast majority of all ANN/AI researchers. It has two phases. The first defines the basic components that might be required for intelligence. In the second, all the components are put together in an enormously complex machine. The AI-GA approach automatically learns how to produce general AI [4]. At present the AI-GA approach is rarely looked at. However, since it can be based much more closely on the brain it is likely to become more popular as our knowledge of the brain improves. Genetic programming could have a big role in AI-GA approaches.

The aim of our work is to find a computational equivalent of the biological neuron and hence general AI. To do this, we propose a simple neural model (DEMANNED) which incorporates both evolution and development. In this two neural programs acting together construct neural networks. The pair build a network from which *multiple* conventional ANNs can be extracted each of which can solve a different computational problem. The model reported here was inspired in part by the developmental method proposed in [22] and particularly by the paper [10]. In previous work, we examined a one spatial dimensional developmental model and this was applied to multiple classification problems only [23, 24]. In [19] we examined a 2D model and investigated the utility of evolving programs that solve problems incrementally. Here, one starts by trying to solve one problem, then after a given number of generations, one tries to solve the first two problems, and so on until eventually one tries to solve all problems at the same time. This was found to be more effective than trying to solve all problems together. Using this strategy we showed that one could evolve developmental programs that could build a neural structure which could achieve reasonable scores on two classification and two control problems at the same time.

In this article we are interested in the utility of using activity dependence (AD). This is where changes in levels of activity between neurons leads to changes in neuron structure and morphology. AD is an extremely important aspect of real brains [25]. Forms of activity dependence have been implemented in the model. These allow neuron (dendrite) health, position and bias (weight) to be affected by signals passing through the neural networks. Activity dependence includes Hebbian-like mechanisms. AD may have a role in alleviating catastrophic interference since the neural network can change when inputs are applied (i.e. during training).

Developmental approaches for building ANNs have long been proposed [13, 30] as one of the important components in an enriched form of artificial evolution called *computational evolution* [2, 34]. In particular, for several decades authors have investigated various ways of implementing and evolving development processes to construct ANNs using a variety of genotype representations at different levels of abstraction. These are reviewed in a recent submission to the Artificial Life journal [20]. However, most previous research in this area has been on small problems and non-standard benchmarks. Also it has not addressed the problem of *multiple problem solving*.

9.2 Multiple Problem Solving ANNs

The main approach to multiple problem solving with ANNs has been to gradually augment ANNs by adding additional neurons or join trained ANNs together via extra connections. This avoids catastrophic forgetting as with each new task a new neural network is established and outputs from neurons in different networks can be shared. 'Constructive neural networks' are traditional ANNs which start with a small network and add neurons incrementally while training error is reduced [6, 7]. Modular ANNs use multiple ANNs each of which has been trained on a sub-problem and these are combined by a human expert [28]. Both of these approaches could be seen as a form of human engineered development. More recent approaches adjust weighted connections *between* trained networks on sub-problems guaranteeing that trained networks on sub-problems are unaltered. Rusu et al. applied their method, called 'progressive neural networks' [27] to three classes of problems: variants of the game of Pong, Atari games and 3D maze problems and Terekhov et al. examined their approach on purpose designed image classification tasks [31]. Aljundi et al. have a set of trained ANNs for each task (experts) and use an additional ANN as a recommender choosing which expert to use for a particular data instance [1]. They evaluated their approach on image classification tasks and video prediction.

Recently a new approach to alleviate catastrophic forgetting in multi-task learning using *neuromodulation* has been proposed. Neuromodulation can help because some neurons in the network can detect which task is currently being performed, and those neurons can turn learning on in the part of the network that performs that task and turn learning off everywhere else in the network. Ellefsen et al. proposed using modular ANNs in which task-specific learning is turned on and off in different modules [5]. Although this reduced catastrophic forgetting, modules specifically dedicated to different problems did not emerge. Velez and Clune implemented diffusion-based neuromodulation in which point sources of a diffusing chemical are placed at specific locations within an ANN [35]. The sources emit diffusing learning signals that correspond to positive and negative feedback for the tasks being learned. On agent-based foraging tasks they were able to create small networks that completely eliminate catastrophic forgetting.

9.3 The Neuron Model

The neural programs are represented and evolved using Cartesian Genetic Programming (CGP) [18, 21] in which the program nodes represent mathematical operations, operating on and returning real-values between -1 and 1. Each primitive function takes up to two inputs, denoted z_0, z_1. The functions are as follows. *Step*: if $z_0 < 0$ then 0 else 1. *Add*: $(z_0 + z_1)/2$. *Sub*: $(z_0 - z_1)/2$. *Mult*: $z_0 z_1$. *Xor*: if the sign of both inputs is the same then the output is -1 else 1. *Istep*: if $z0$ is negative, output is 1 else output is 0. These functions were found to be effective in previous work. The

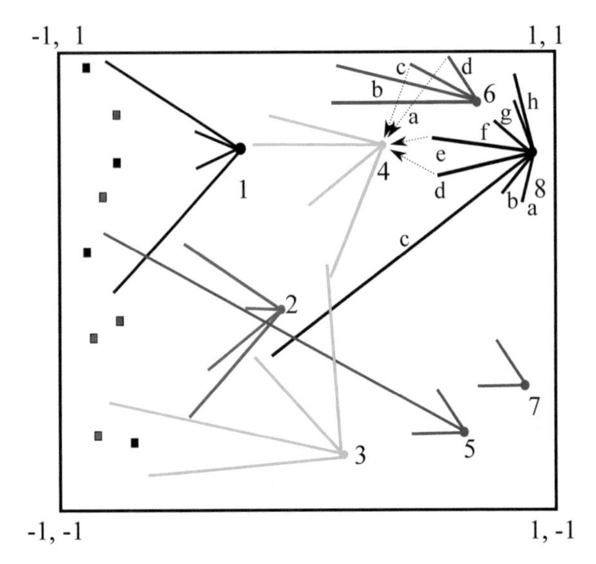

Fig. 9.1 A fictitious brain with 8 neurons solving three problems. Two output neurons are devoted to computational problem one (black), three are devoted to problem two (blue) and one is devoted to problem three (red). There are also two non-output neurons (green). Problem one has three inputs (black squares on left), problem two has two inputs (blue) and problem three has three inputs (red). Inside each neuron and dendrite is a CGP-encoded program (see Fig. 9.2)

programs read variables associated with neurons and dendrites and produce outputs which are used to update those variables.

To illustrate the model we discuss a fictitious brain example as shown in Fig. 9.1. The dendrites are assumed to connect to the nearest neurons or inputs on the left of the parent neuron. Neurons only have a soma and a number of dendrites. Output neurons are dedicated to their corresponding computational problem. However, it is possible that some non-output neurons can be shared between different problems. In Fig. 9.1 we can see that dendrites c and d belong to neuron 6 and will connect to neuron 4 (see dotted arrows). Also dendrites e and d of neuron 8 will also connect to neuron 4. Thus in this example non-output neuron 4 is shared between computational problems one and two.

The inputs and outputs of evolved programs are shown in Fig. 9.2. When the evolved soma and dendrite programs are executed, neurons can move, change, die replicate, grow more dendrites and their dendrites can also change, replicate or die. We refer to the collection of neurons as the *brain*. Neurons and dendrites are confined to the unit square and all neural variables can only take values between -1.0 and 1.0. There are two kinds of neuron: output and non-output. Every output required by each computational problem has a dedicated *output neuron*. The other neurons are internal and are not used to provide outputs from the brain. We refer to these as *non-output neurons*.

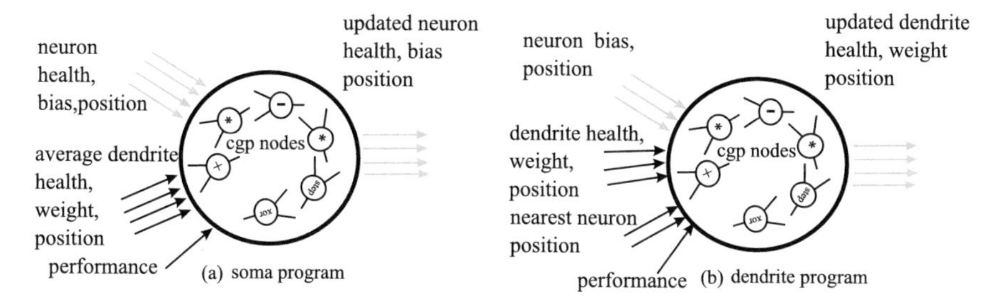

Fig. 9.2 The model of a developmental neuron

9.3.1 Soma Program Inputs and Outputs

The soma program can read up to nine variables. The first four are the neuron variables: x and y position, health and bias. Bias refers to an input to the neuron activation function which is added to the weighted sum of inputs (i.e. it is unweighted). The soma program can also be supplied with averages of properties of its dendritic tree: x and y position, weight and health. Finally, the soma program can read a reward signal which is related to how well the brain is performing on the suite of computational problems. Non-output neurons receive a reward signal equal to the fitness score. Output neurons receive the normalised fitness score corresponding to the problem the output neuron belongs to.

The soma program has four outputs: health updater (sh_u), bias updater (sb_u), and x and y position updater (sx_u and sy_u). The evolved soma program reads its ten inputs and outputs these four soma output update variables. These decide how the actual soma corresponding variables will be updated. The way this is done is as follows. If any soma updater variable is greater (less) than zero, the corresponding soma variable is incremented (decremented) by a user-defined amount (later referred to as $delta$). In the case of soma health, there is a further step. If it falls below the user-defined death threshold, NH_{death}, then the neuron will die and not be present in the updated brain. Alternatively, if it happens to be above the user-defined neuron birth threshold, NH_{birth}, then the parent neuron will replicate and an additional neuron will appear in the brain (near to the parent). In this way, the soma evolved programs can change the health, bias or position of the soma and whether the neuron will die, or replicate.

9.3.2 Dendrite Program Inputs and Outputs

The dendrite program is executed inside every dendrite. It can read up to nine variables. The first three are the parent neuron variables: position (x and y) and bias. The parent neuron health was not supplied to the dendrite program. Initial experiments indicated that this produced superior results. The next four are the dendrite variables:

x and y position, weight and health. The dendrite program also can read the x and y position of the nearest neuron to the dendrite position.

The dendrite program also has four outputs: health updater (dh_u), weight updater (dw_u), and x and y position updater (dx_u and dy_u). The evolved dendrite program reads its nine inputs and outputs these four dendrite output update variables. These decide how the actual dendrite corresponding variables will be updated. The way this is done is as follows. If any dendrite updater variable is greater (less) than zero, the corresponding dendrite variable is incremented (decremented) by a user-defined amount (later referred to as $delta$). If the *parent* neuron health is above a user-defined dendrite health birth threshold, DH_{birth}, then the parent neuron will gain a new dendrite. Dendrites die if the *dendrite health* is below (above) a user-defined threshold, DH_{death}.

9.3.3 Developing the Brain and Evaluating the Fitness

The algorithm used for training and developing the ANNs is given in Algorithm 3. The brain is always initialised with at least as many neurons as the maximum number of outputs over all computational problems. Note, all problem outputs are represented by a unique neuron dedicated to the particular output. Output neurons can change, but not die or replicate as the number of output neurons is fixed by the choice of computational problems. The number of developmental steps are defined by the parameters, NDS_{pre} and NDS_{whi}. The 'pre' learning phase is an initial phase of development where the brain is not tested in any way (lines 5–7). In the 'while' phase the brain is assessed and provides feedback to the developmental process (lines 10–12).

Lines 9–30 form the *epoch learning* loop. This loop repeats the entire training developmental process (the 'while loop') for a number of epochs, N_{ep}. The purpose of learning epochs is to allow us to direct evolution to produce a pair of programs that cause the developing ANN to learn. The neural programs can read the performance of the brain at the previous learning epoch. The learning loop only continues while the training accuracy does not decrease (lines 25–29). If it does, the algorithm stops and returns the training score of the previous epoch.

Note that at each epoch, a performance value is determined corresponding to each individual benchmark problem and this an input to the soma and dendrite programs for *output* neurons. If a neuron is not an output neuron then the average fitness over all problems at the previous epoch, is given as an input to the soma and dendrite programs. The performance signal is intended to act as a reward to the developmental process, triggering changes in the brain when necessary.

Algorithm 3 Fitness algorithm.

```
1: function FITNESS
2:     Initialise brain
3:     Load 'pre' development parameters
4:     PrevFitness = 0
5:     for NDS_pre times do
6:         Run soma/dendrite programs to update brain
7:     end for
8:     Load 'while' developmental parameters
9:     for epoch = 1 to N_ep do
10:        for NDS_whi times do
11:            Run evolved programs to update brain
12:        end for
13:        TotFitness = 0
14:        for p = 1 to NumBenchmarkProblems do
15:            Extract ANN
16:            Fitness(p) = 0
17:            for NT(p) training cases do
18:                Make activity-dependent changes
19:                Fitness(p) = Fitness(p) + FitInstance
20:            end for
21:            Fitness(p) = Fitness(p)/NT(p)
22:            TotFitness = TotFitness + Fitness(p)
23:        end for
24:        TotFitness = TotFitness/NumBenchmarkProblems
25:        if TotFitness < PrevFitness then
26:            Break
27:        else
28:            PrevFitness = TotFitness
29:        end if
30:    end for
31:    Return PrevFitness
32: end function
```

9.3.4 Extracting Conventional ANNs from the Brain

Conventional ANNs are extracted from the developed brain (line 15 in Algorithm 3). Input data is supplied at fixed spatial locations unique to each problem. First, the maximum number of inputs in all the computational problems used in fitness evaluation is determined. This number of inputs is used to assign inputs fixed random positions. When data for a particular computational problem is presented, any inputs that are undefined for that problem are assumed to be zero.

The next phase is to go through all dendrites of the neurons to determine which inputs or neurons they connect to. To generate a valid neural network we assume that dendrites are automatically connected to the nearest neuron or input on the left. We refer to this as *snapping*. Since, the dendrite position can be on the right of the parent neuron before extracting ANNs, it is reflected back from the parent position. The dendrites of non-output neurons are allowed to connect to either inputs or other non-output neurons on their left. However, output neurons are only allowed to connect to *non-output* neurons on their left. Although, it is not desirable for the dendrites of output neurons to be connected directly to inputs, when output neurons move, they may only have inputs on their left. $NT(p)$ is the number of training cases for each computational problem, p. The extracted ANNs use the hyperbolic tangent activation function. A number of alternative activation functions were examined (e.g. sigmoid, rectilinear) and hyperbolic tangent seemed to be the most effective.

In Algorithm 3 development in the epoch learning loop happens *outside* the benchmark problem loop (i.e. lines 14–23). However, in principle it could be placed within it (immediately after line 14). This would allow brain development during problem solving. This could help the brain to develop differently according to the problem currently being solved. Such an option has been implemented but as yet has not shown performance advantages.

9.3.5 Activity Dependence

Step 18 in Algorithm 3 allows the strength of the signal to cause changes in the brain (activity dependence). We have implemented AD mechanisms to affect neuron health and bias and dendrite weight, health and position. We have also implemented a form of non-temporal Hebbian adjustment to dendrite weight. These mechanisms are shown in Algorithm 4. W_{ij} is the weight of dendrite j of neuron i *in the extracted ANN*. $Brain_{ij}(v)$ is variable v of dendrite j of neuron i (i.e. v can be health, weight, or position). $BrainW_{ij}$ is the weight of dendrite j of neuron i in the brain. DS_{ij} denotes the signal passing through dendrite j belonging to neuron i. In line 6 the difference between the absolute value of the dendrite signal and a user defined threshold (θ_v) is calculated. Either the variables, weight, health or position of the dendrite in the brain is then *reduced* in magnitude using the user-defined increment corresponding to the chosen variable, δ_{dv}^{act}). This is a homeostatic mechanism [3] where with large signals, neural variables reduce to maintain homeostasis. For brevity we have only shown one activity-dependent adjustment. The user can choose to adjust any or all of the three dendrite variables in this way. The weighted sum of signals over all dendrites belonging to a neuron is accumulated and a bias B_i is added (line 13). The neuron signal, NS_i, is then calculated using a user-defined slope parameter, α (line 14). Then if AD is chosen for a neuron (lines 39–44) brain adjustments take place according to whether the magnitude of the neuron signal, $| NS_i |$ exceeds a user-defined theshold, θ_v. In this article, we are particularly interested in AD dendrite weight. The neuron health or position could also be adjusted in a similar manner.

Algorithm 4 Signal propagation and activity dependence.

1: **for** neuron i **do**
2: $W_{sum} = 0$
3: **for** dendrite ij **do**
4: $W_{sum} = W_{sum} + W_{ij} \times DS_{ij}$
5: **if** Dendrite activity dependence **then**
6: $D =\mid DS_{ij} \mid - \theta_v$
7: **if** $D \geq 0$ **then**
8: $Brain_{ij}(v) = Brain_{ij}(v) - \delta_{dv}^{act}$
9: Bound $Brain_{ij}(v)$
10: **end if**
11: **end if**
12: **end for**
13: $W_{sum} = W_{sum} + B_i$
14: $NS_i = \tanh(\alpha W_{sum})$
15: **if** Hebbian learning **then**
16: **if** $NS_i > \theta_{Hebb}$ **then**
17: $N_{high} = 1$
18: **end if**
19: **if** $NS_i < -\theta_{Hebb}$ **then**
20: $N_{low} = 1$
21: **end if**
22: **for** dendrite ij **do**
23: **if** $DS_{ij} > \theta_{Hebb}$ **then**
24: $D_{high} = 1$
25: **end if**
26: **if** $DS_{ij} < -\theta_{Hebb}$ **then**
27: $D_{low} = 1$
28: **end if**
29: $Both_{high} = N_{high}$ AND D_{high}
30: $Both_{low} = N_{low}$ AND D_{low}
31: **if** $Both_{high}$ OR $Both_{low}$ **then**
32: $BrainW_{ij} = BrainW_{ij} + \delta_{inc}^{Hebb}$
33: Bound $BrainW_{ij}$
34: **else**
35: $BrainW_{ij} = BrainW_{ij} \times \delta_{mult}^{Hebb}$
36: **end if**
37: **end for**
38: **end if**
39: **if** Neuron activity dependence **then**
40: $D =\mid NS_i \mid - \theta_v$
41: **if** $D \geq 0$ **then**
42: $Brain_i(v) = Brain_i(v) - \delta_{sv}^{act}$
43: Bound $Brain_i(v)$
44: **end if**
45: **end if**
46: **end for**

Lines 15–38 are concerned with Hebbian-like learning. Here, if the magnitude of the signal passing along the dendrite and the output of the parent neuron both exceed a threshold (i.e. they agree), then the weight of the dendrite is increased using the user-defined increment, δ_{inc}^{Hebb}. However, if only one exceeds a threshold (i.e. they disagree) then the weight is decreased (using the user-defined multiplier, δ_{mult}^{Hebb}). It should be noted that the model has many parameters many of which are thresholds and allowed increments on neural variables. The frequency of AD changes can be controlled via the corresponding threshold, θ_v.

9.3.6 Model Parameters

The model has a large number of user-defined parameters (Table 9.1). It is hoped that as the model develops some can be removed or assume default values, however, at present the approach has been to create a neuron model that is as general as possible and with the least number of assumptions. The initial number of non-output neurons, can be chosen by the user and is denoted, N_{init}. In addition, each output for each computational problem has an output neuron. The total number of neurons allowed in the network is bounded between a user-defined lower (upper) bound NN_{min} (NN_{max}). The number of dendrites each neuron can have is bounded by user-defined lower (upper) bounds denoted by DN_{min} (DN_{max}). These parameters ensure that the number of neurons and connections per neuron remain in well-defined bounds, so that a network can not eliminate itself or grow too large.

If the health of a neuron falls below (exceeds) a user-defined threshold, NH_{death} (NH_{birth}) the neuron will be deleted (replicated). Likewise, dendrites are subject to user defined health thresholds, DH_{death} (NH_{birth}) which determine whether the dendrite will be deleted or a new one will be created. Actually, to determine dendrite birth the parent *neuron* health is compared with DH_{birth}. This choice was made to prevent the potentially very rapid growth of dendrite numbers.

When neurons are initialized their health and bias are given random values between -1 and 1. All neurons are initialized with ND_{init} dendrites. The dendrites variables, health and weight are initialized with random values between -1 and 1. They are given randomly chosen x and y-positions between 0 and 1. When neurons are born they are given ND_{init} dendrites. Finally, the neural activation function has a slope constant given by α.

Newly born neurons are given a health equal to one, a bias of zero, and ND_{init} dendrites. They are placed above and to the right of the parent neuron, by adding a small increment, MN_{inc} to the parent's x and y position. Their dendrites are given weight equal to zero and a health equal to one. The x and y positions of the dendrites are set to zero. When a neuron decides to create a new individual dendrite it is given a weight and health equal to one and x and y-positions equal to 0.8 of the parent neuron x and y-position. There are many possible choices for these parameters when neurons and dendrites are born. In preliminary empirical investigations these choices were found to work well.

Table 9.1 Table of neural model constants, their meanings and chosen values

Symbol	Meaning	Value
$NN_{min}(NN_{max})$	Min. (Max.) allowed number of neurons	0 (30)
N_{init}	Initial number of non-output neurons	6
$DN_{min}(DN_{max})$	Min. (Max.) number of dendrites per neuron	1 (60)
ND_{init}	Initial number of dendrites per neuron	5
$NH_{death}^{pre}(NH_{birth}^{pre})$	Neuron health thresholds for death (birth)	−0.6 (0.2)
$DH_{death}^{pre}(DH_{birth}^{pre})$	Dendrite health thresholds for death (birth)	−0.6 (0.2)
$NH_{death}^{whi}(NH_{birth}^{whi})$	Neuron health thresholds for death (birth)	−0.4 (0.1)
$DH_{death}^{whi}(DH_{birth}^{whi})$	Dendrite health thresholds for death (birth)	−0.64 (−0.61)
δ_{nh}	Neuron health increment: pre (while)	0.2 (0.1)
δ_{np}	Neuron position increment: pre (while)	0.1 (0.1)
δ_{nb}	Neuron bias increment: pre (while)	0.1 (0.1)
δ_{dh}	Dendrite health increment: pre (while)	0.2 (0.1)
δ_{dp}	Dendrite position increment pre (while)	0.1 (0.1)
δ_{dw}	Dendrite weight increment: pre (while)	0.1 (0.1)
NDS_{pre}	Number of developmental steps	6
NDS_{whi}	Number of 'while' developmental steps	1
N_{ep}	Number of learning epochs	8
MN_{inc}	Move neuron increment if collision	0.0001
I_u	Max. initial input position	−0.6
O_l	Neuron output position on x-axis	1
α	Hyperbolic tangent exponent constant	1.5
θ_{dw}^{AD}	Threshold for AD weight change	0.7
δ_{dw}^{AD}	AD weight change increment	0.08

In some cases, neurons will collide with other neurons (by occupying a small interval around an existing neuron) and the neuron has to be moved by a certain increment until no more collisions take place. This increment is given by MN_{inc}.

The x positions of data inputs to the brain are given fixed random values between -1 and $-1 + I_u$ while the y-positions take randomly chosen values between -1 and 1. The output neurons for all problems are initially placed at x-position O_l and their positions on the y-axis are uniformly distributed between -1 and 1. However, output neurons as with other neurons can move according to the neuron program. All neurons are marked as to whether they provide an external output or not. In the initial network there are N_{init} non-output neurons and N_o output neurons, where N_o denotes the number of outputs required by the computational problem being solved. When solving a particular problem, output data is read from only those output neurons corresponding to the chosen problem (the remaining output neurons are ignored). Note, non-output neurons are not allowed to connect to output neurons and output neurons can only connect to non-output neurons or inputs.

The chosen experimental parameters for this study are also shown in Table 9.1. After much experimentation these were found to work quite well. The values found in the while phase $DH_{death} = -0.64$ and $DH_{birth} = -0.61$ are quite surprising as learning is taking place while there is a high probability of dendritic change (birth or death).

9.4 Experiments

In this work, we look to simultaneously solve collections of problems chosen from two standard classification problems, diabetes (D) and glass (G) and two reinforcement learning problems , double-pole balancing (DP) [33] and ball throwing (BT) [12, 33]. The definitions of the classification problems are available in the UCI repository.[1] D has 8 real attributes and two Boolean outputs. G has 9 real attributes and six Boolean outputs. The Boolean class is decided by whichever ANN output is the greater.

In BT, the aim is to design a controller which throws a ball as far as possible. There are two inputs, the arm angle from vertical and the angular velocity of the arm. It has two outputs, the applied torque to the arm and an output which decides when to release the ball. The system is simulated for a maximum of 10,000 time steps. The maximum distance the ball can be thrown is can be determined through simulation and has a value of approximately, 10.202 m. BT is considered solved when the thrown distance greater than or equal to 9.5 m (fitness = 0.9312). It has a strong sub-optimal fitness value where the fitness is 0.546 this corresponds to the maximum possible distance that the ball can be thrown when the arm only swings forward, whereas to achieve maximum distance, one needs to swing the arm backwards so that it picks up speed due to gravity before torque is applied.

[1] https://archive.ics.uci.edu/ml/datasets.html.

Table 9.2 Performance for problem pairs with activity dependence versus no activity dependence

Problem pair	AD	No AD
DP/BT	0.3869	0.3287
DP/G	0.4230	0. 4016
D/G	0.6506	0.6482
BT/G	0.5585	0.5768

In DP, the task is to balance two poles on a moveable cart on a limited track by applying a horizontal force to the cart. The inputs to the controller are the position and velocity of the cart and the angle and angular velocity of the pole(s). So there are six inputs. The single output is the force applied to the cart. The system is simulated for a maximum of 100,000 time steps. It is solved if both poles are balanced to within certain limits for this number of steps. The fitness for the DP problem is the fractional number of simulation steps that the poles remain balanced so the fitness is fractional while the fitness for the ball throwing problem is a continuous floating point value.

The Wilcoxon Ranked-Sum test (WRS) was used to assess the statistical difference between pairs of experiments [17]. In this test, the null hypothesis is that the results over the multiple runs for the two different experimental conditions are drawn from the same distribution and have the same median. If there is a statistically significant difference between the two then null hypothesis is false with a degree of certainty which depends on the smallness of a calculated statistic called a p-value. However, in the WRS before interpreting the p-value one needs to calculate another statistic called Wilcoxon's W value. This value needs to be compared with calculated values which depend on the number of samples in each experiment. Results are statistically significant when the calculated W-value is less than or equal to certain critical values for W_c [36]. In all our experiments $W_c = 38$. This is available in standard tables of values dependent on the number of paired samples (20 in our case) and the p-value bounds for significance.

The genotype length was chosen to be 600 nodes. Goldman mutation [9] was used which carries out random point mutation until an active gene is changed. Twenty non-incremental evolutionary runs of a 1+5-ES were used using 20,000 generations. In a series of experiments we compared the effectiveness of solving pairs of problems with AD weight versus solving them without AD weight. The results are shown in Fig. 9.2. It turned out that all W-values were larger than the critical value, so the statistical differences between the two scenarios were not significant. Other experiments were conducted with both Hebbian and activity-dependent dendrite position and once again there appeared to be no statistical differences between allowing activity dependence and disallowing it.

9.5 Discussion and Further Work

The findings regarding the usefulness of activity dependence were disappointing. However, there are other ways that activity dependence could be implemented. Average activity could be calculated over each problem data set and this could be given as an input to the evolved programs. This could allow evolution to respond to activity levels. In the AD implemented in this article, information about neural activity was not supplied to evolved programs but rather neural activity was calculated and neural variables adjusted *after* the ANNs were extracted from the developed brain. Ideally, evolved neural programs would be executed during training, however this would be computationally prohibitive. In real brains, neurons and dendrites are all running in parallel.

There are many small details in the model particularly relating to the birth of neurons and dendrites. Empirical investigations to ascertain the most suitable value of the many parameters have not been exhaustive but rather based on small semi-informal experiments. It is therefore possible that there are much more suitable parameters.

Although we were able to evolve a computational brain that can solve multiple machine learning problems reasonably well, our attempts were greatly hindered by interference. Activity dependence could in principle alleviate catastrophic forgetting as the networks could change depending on the problem input data. How does natural evolution find improvements in the performance of systems without the deterioration of already evolved systems? For computational reasons the maximum number of neurons we have used in the experiments has been very small (30). One would imagine with much larger numbers of neurons that it would be easier for evolution to develop sub-networks of neurons without interfering with already successful ones. It could also be simply that to achieve non-interfering development takes much more evolutionary time. Allowing development to happen within the problem loop (as discussed earlier in Sect. 9.3.3) needs to be more thoroughly investigated.

Snapping needs to be investigated further. At present dendrites snap to their nearest neurons to establish a connection. However, snapping could be made more local, so that dendrites only snap when the distance between them and the neuron is less than a certain bound. Unconnected dendrites would have to be ignored when extracting ANNs.

The idea behind epoch learning was to allow development to take place until the brain stopped improving. The hope is that by evolving developmental programs over sufficient numbers of epochs would encourage generality and result in a self-improving brain. Environmental feedback (training fitness score) were introduced to give signals to the developing brain that would allow it to improve. In general, we found that highest performing developmental programs used few epochs. This remains puzzling and needs further investigation.

More thought needs to be given as to what internal reward signals need to be given to the brain. Biological brains have networks of neurons that *recognise* what the problem is so that the appropriate rules and actions can be applied. In other words, the mechanism for selecting relevant inputs is highly complex.

Generality in learning is commonly evaluated using unseen data sets, however in multiple problem solving methods one could test general learning by presenting a new problem of the same type encountered in training. This will be investigated in the future. We have looked at supplying problems to solve both sequentially and simultaneously. However, perhaps problems should be presented randomly.

We have evaluated our developmental approach on standard benchmarks in machine learning. This was deliberate so that comparisons could be made with other techniques. However, it might be better to create new suites of simpler problems for developmental methods. Also, solving much larger numbers of problems might bias evolution toward general learning behaviour.

References

1. Aljundi, R., Chakravarty, P., Tuytelaars, T.: Expert gate: Lifelong learning with a network of experts. CoRR, abs/1611.06194 **2** (2016)
2. Banzhaf, W., Beslon, G., Christensen, S., Foster, J.A., Képès, F., Lefort, V., Miller, J.F., Radman, M., Ramsden, J.J.: From artificial evolution to computational evolution: A research agenda. Nat. Rev. Genet. **7**, 729–735 (2006)
3. Butz, M., Wörgötter, F., van Ooyen, A.: Activity-dependent structural plasticity. Brain Res. Rev. **60**(2), 287–305 (2009)
4. Clune, J.: AI-GAs: AI-generating algorithms, an alternate paradigm for producing general artificial intelligence (2020). arXiv:e1905.10985
5. Ellefsen, K., Mouret, J.B., Clune, J.: Neural modularity helps organisms evolve to learn new skills without forgetting old skills. PLoS Comput. Biol. **11**(4:e1004128) (2015)
6. Fahlman, S.E., Lebiere, C.: The cascade-correlation learning architecture. In: Advances in Neural Information Processing Systems, pp. 524–532 (1990)
7. Franco, L., Jerez, J.M.: Constructive Neural Networks, vol. 258. Springer (2009)
8. French, R.M.: Catastrophic forgetting in connectionist networks: causes, consequences and solutions. Trends Cognit. Sci. **3**(4), 128–135 (1999)
9. Goldman, B.W., Punch, W.F.: Analysis of cartesian genetic programmings evolutionary mechanisms. IEEE Trans. Evolut. Comput. **19**, 359–373 (2015)
10. Khan, G.M., Miller, J.F., Halliday, D.M.: Evolution of cartesian genetic programs for development of learning neural architecture. Evol. Comput. **19**(3), 469–523 (2011)
11. Kleim, J., Napper, R., Swain, R., Armstrong, K., Jones, T., Greenough, W.: Selective synaptic plasticity within the cerebellar cortex following complex motor skill learning. Neurobiol. Learn. Mem. **69**, 274–289 (1998)
12. Koutník, J., Gomez, F., Schmidhüber, J.: Evolving neural networks in compressed weight space. In: Proceedings of the Genetic and Evolutionary Computation Conference, pp. 619–626 (2010)
13. Kumar, S., Bentley, P. (eds.): On Growth, Form and Computers. Academic (2003)
14. Maguire, E.A., Gadian, D.G., Johnsrude, I.S., Good, C.D., Ashburner, J., Frackowiak, R.S.J., Frith, C.D.: Navigation-related structural change in the hippocampi of taxi drivers. PNAS **97**, 4398–4403 (2000)
15. McCloskey, M., Cohen, N.: Catastrophic interference in connectionist networks: the sequential learning problem. Psychol. Learn. Motivat. **24**, 109–165 (1989)
16. McCulloch, Pitts, W.: A logical calculus of the ideas immanent in nervous activity. Bull. Math. Biophys.**5**, 115–133 (1943)
17. McDonald, J.H.: Handbook of Biological Statistics, 3 edn. Sparky House Publishing (2014)

18. Miller, J.F. (ed.): Cartesian Genetic Programming. Springer (2011)
19. Miller, J.F.: Evolving developmental neural networks to solve multiple problems. In: Proceedings of ALIFE-2020, pp. 473–482 (2020)
20. Miller, J.F.: DEMANNED: Designing multiple ANNs via evolved developmental neurons. Artificial Life (2021), submitted
21. Miller, J.F., Thomson, P.: Cartesian genetic programming. In: Proceedings of European Conference on Genetic Programming, LNCS, vol. 10802, pp. 121–132 (2000)
22. Miller, J.F., Thomson, P.: A Developmental Method for Growing Graphs and Circuits. In: Proceedings of International Conference on Evolvable Systems, LNCS, vol. 2606, pp. 93–104 (2003)
23. Miller, J.F., Wilson, D.G., Cussat-Blanc, S.: Evolving developmental programs that build neural networks for solving multiple problems. In: Genetic Programming Theory and Practice XVI (pp. 137–178 (2019)
24. Miller, J.F., Wilson, D.G., Cussat-Blanc, S.: Evolving programs to build artificial neural networks. In: From Astrophysics to Unconventional Computation, pp. 23–71. Springer International Publishing (2020)
25. Ooyen, A.V. (ed.): Modeling Neural Development. MIT Press (2003)
26. Ratcliff, R.: Connectionist models of recognition and memory: constraints imposed by learning and forgetting functions. Psychol. Rev. **97**, 205–308 (1990)
27. Rusu, A.A., Rabinowitz, N.C., Desjardins, G., Soyer, H., Kirkpatrick, J., Kavukcuoglu, K., Pascanu, R., Hadsell, R.: Progressive neural networks (2016). arXiv:1606.04671
28. Sharkey, A.J.: Combining Artificial Neural Nets: Ensemble and Modular Multi-net Systems. Springer Science & Business Media (2012)
29. Smythies, J.: The Dynamic Neuron. MIT Press (2002)
30. Stanley, K.O., Miikkulainen, R.: A taxonomy for artificial embryogeny. Artif. Life **9**(2), 93–130 (2003)
31. Terekhov, A.V., Montone, G., O'Regan, J.K.: Knowledge transfer in deep block-modular neural networks. In: Conference on Biomimetic and Biohybrid Systems, pp. 268–279. Springer (2015)
32. Tramontin, A.D., Brenowitz, E.: Seasonal plasticity in the adult brain. Trends Neurosci. **23**, 251–258 (2000)
33. Turner, A.J.: Evolving Artificial Neural Networks using Cartesian Genetic Programming. Ph.D. thesis, Department of Electronic Engineering, University of York (2017). http://etheses.whiterose.ac.uk/12035/
34. Vaario, J.: From evolutionary computation to computational evolution. Informatica **18**, 417–434 (1994)
35. Velez, R., Clune, J.: Diffusion-based neuromodulation can eliminate catastrophic forgetting in simple neural networks. PLOS One **12**(11:e0187736) (2017)
36. Zar, J.H.: Biostatistical Analysis, 2nd edn. Prentice Hall (1984)

Chapter 10
Evolving and Analyzing Modularity with GLEAM (Genetic Learning by Extraction and Absorption of Modules)

Anil Kumar Saini and Lee Spector

Abstract General methods for the evolution of programs with modular structure have long been sought by genetic programming researchers, in part because modularity has long been considered to be essential, or at least helpful, for human programmers when they develop large-scale software projects. Multiple efforts have been made in this direction, and while success has been demonstrated in specific contexts, no general scheme has yet been demonstrated to provide benefits for evolutionary program synthesis that are similar in generality and significance to those provided by modularity in human software engineering. In this chapter, we present and analyze a new framework for the study of the evolution of modularity, called GLEAM (Genetic Learning by Extraction and Absorption of Modules). GLEAM's flexible architecture and tunable parameters allow researchers to test different methods related to the generation, propagation, and use of modules in genetic programming.

10.1 Introduction

Genetic programming systems have been shown to be capable of synthesizing programs that make use of multiple data-types, conditionals, loops, and other control structures that, for human programmers, support the development of complex programs. They can successfully solve many problems from the introductory programming textbooks [4]. To solve more complex problems, however, like many that are routinely solved by human programmers, additional breakthroughs may be required with respect to the evolution of modularity. Despite multiple efforts in this direction, achieving scalability through modularity remains one of the open issues in the field of genetic programming [10].

A. K. Saini (✉)
University of Massachusetts Amherst, Amherst, MA, USA
e-mail: aks@cs.umass.edu

L. Spector
Amherst College, University of Massachusetts Amherst, Amherst, MA, USA
e-mail: lspector@amherst.edu

W. Banzhaf et al. (eds.), *Genetic Programming Theory and Practice XVIII*,
Genetic and Evolutionary Computation,
https://doi.org/10.1007/978-981-16-8113-4_10

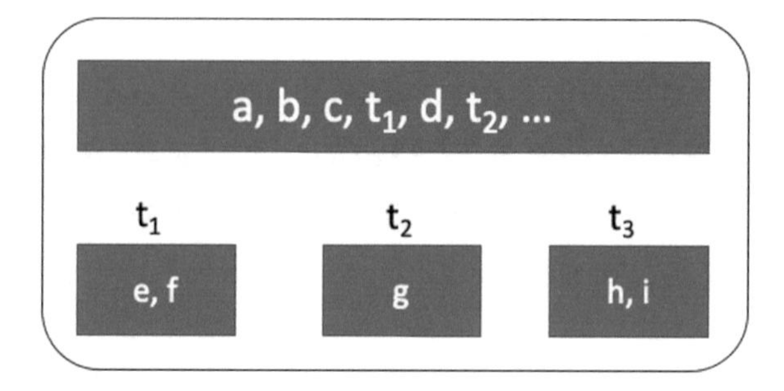

Fig. 10.1 An individual in GLEAM. The first part shows the program, and the the second part shows the modules labeled with tags. The letters *a*, *b*, *c*, etc. denote the regular instruction, whereas t_1, t_2, etc., denote module references

Many of the techniques proposed to encourage modularity in evolving programs are designed to work only for a handful of modules, and it is not clear that these techniques will be able to produce large-scale software that involves dozens or more modules. In this chapter, we present and analyze a new framework called GLEAM (Genetic Learning by Extraction and Absorption of Modules), in which an evolving program can make use of a local library of modules that is propagated from generation to generation (with possible variation) along with the program's code. Programs can call modules, and modules can also call each other. Figure 10.1 illustrates a typical evolving individual in GLEAM.

The framework presented here does not specify how module arguments and return values and their types are handled. Details of argument handling will depend on the underlying genetic programming system; below, we spell out the way in which this is handled in our own experiments.

After every generation, the library is updated in the following way: modules that are not being used by the individual may be replaced by modules extracted from the individual itself or elsewhere, modules may be mutated, and module references may be replaced by their corresponding code segments (which we call "absorption" of the module). New modules are typically extracted from the main program or other modules associated with the same program, but under certain experimental conditions they may instead be taken from other individuals or any other external source.

Below, after a brief review of related work, we present the GLEAM framework in more detail and describe the results of experiments with GLEAM on benchmark software synthesis problems. We analyze the effectiveness of GLEAM for evolving modular programs, and the improvement in software synthesis performance relative to experiments in which GLEAM is not used. We also briefly describe additional experiments that illustrate the use of GLEAM as a platform for testing different methods for evolving and using modules in genetic programming.

10.2 Evolving Modules in Genetic Programming

Ever since the introduction of Automatically Defined Functions (ADFs) by John Koza [8], multiple attempts have been made over the years to evolve modular programs that can solve complex problems. Although different frameworks define modularity in different ways, in this section, we restrict our discussion to those frameworks where evolving individuals have access to *labeled* modules.

These modularity-inducing systems can be grouped into two categories. In the systems in the first category, which we call Single-entry-point systems, an evolving individual is composed of a main program and a set of modules. These modules can be local to an individual or shared by all individuals in the population. Execution always starts from the main program, which can call other modules during execution. Modules can also call each other. In the second category, which we call Multiple-entry-points systems, an evolving individual is made up of modules. There is no concept of a main program as such, and the execution can start from any of the modules based on the signal from the environment.

Single-entry-point Systems. John Koza's Automatically Defined Functions (ADFs), one of the first approaches in this direction, were able to improve the probability of success on a number of problems such as parity functions. However, they were not very flexible in the sense that the form of the functions, including their name, arguments, etc. had to be defined in advance. To remedy this, various modifications have come up, including but limited to, Architecture Altering Operations [7], Module Acquisition [1], Automatically Defined Macros [11], etc.

Grammatical Evolution (GE) using modules is another technique in this category. First, all the subtrees in the individuals in the population are assessed on their usefulness in their respective program trees, and the better performing ones are considered to be modules. This is called *module identification* [15]. Next, the underlying grammar is modified so that individuals in the future generations are able to use such modules. Tag-based modules [13], whereby programs can label code segments with integer tags and later refer to them during execution, have shown limited success in problems like the lawnmower problem and the obstacle-avoiding robot problem. Another technique that encourages modularity in the evolving programs by reusing and evolving modules is Embedded Cartesian Genetic Programming (ECGP) [16].

Some work has also been done to use a set of modules between different runs for a single problem. For example, a Run Transferable Library [5], which is a collection of functions known as Tag Addressable Functions (TAFs), is used and updated in one run and transferred to the next one for a certain number of runs for the same or similar problems. All the individuals in the population are free to use the modules in this global library.

Multiple-entry-points Systems. In SignalGP [9], a program is a set of functions that can be accessed by their identifiers called tags. Events in the environment also contain tags and can trigger functions with appropriate tags in the program. Tangled program graphs (TPG) [6] use two distinct populations, one for teams and the other for programs. A team is a collection of pointers to programs, which are executed to calculate the fitness of the team. Here, teams act as individuals and programs as modules.

The GLEAM framework that we discuss in this chapter primarily focuses on the systems in the first category since it assumes that an evolving individual is made up of a main program and a set of labeled modules.

10.3 GLEAM

In this section, we describe various aspects of the GLEAM framework. Although the description has been kept as general as possible, some details on how GLEAM has been implemented in our experiments are also given.

Each individual in a GLEAM population has a main program and a local library containing a number of modules. Only the main program is executed for error testing. Modules are not executed directly, but called by the main program or by other modules during execution. Modules are identified and called using integer tags.

Libraries are updated each generation, ideally in a way that will cause useful modules to be retained and useless modules to be replaced by new ones over the course of evolution. We define the usefulness of modules in a simple way: those modules which are being referenced by the program directly or indirectly are considered useful. Those that cannot be reached by chains of references starting in the main program are considered useless.

GLEAM is a general-purpose technique; it can be implemented with most, if not all, of the existing genetic programming systems. The only requirement for the underlying system is that the modules used by the evolving individuals in the system have *labels* that can be used to call them. For the sake of concreteness, the description of GLEAM in this section assumes the use of linear genome representations, but we provide some suggestions for implementation in tree-based genetic programming systems as well.

10.3.1 Initializing the Library

Modules are initialized in the same way as the program itself, using the same instructions that are available to the program.

The number of modules in the library of an individual can be set to any size. In the version of GLEAM that we use for the preliminary experiments described in this chapter, each individual library contains a small fixed number of modules (10). However, we consider this to be a special case of the more general GLEAM architecture, in which the number of modules may be increased over evolutionary time, with a specified number of modules being added every time a specified number of generations passes. More research will be required to determine the effects of various schedules for increasing the number of modules, and to understand the behavior of the system with much larger numbers of modules.

Within the context of a fixed, small number of modules, we found in exploratory experiments that the precise number makes little difference, and we picked 10 as the number of modules for the experiments here simply because it is a round number.

10.3.2 Referencing the Modules

Whenever a module is called, it is provided the current state of the calling program. During execution, it can change that state, and that state later on gets returned to the calling program after the module has been executed. Restrictions on the amount and type of information passed to the modules, and returned from the module calls, could be implemented for the sake of specific experiments, but we do not do so here. Instead, we leave it to the underlying genetic programming systems to deal with it.

Calling a module requires the use of special referencing instructions. For the experiments described here, referencing instructions are implemented in the following way. In order to refer to a particular module with identifier i (also called its tag), the program uses the instruction `tagged_i`. This is similar to the procedure adopted in [13]. To generate tag references in the program, a special function `tagged_erc_limit` is used. It inserts `tagged_j` in the program, where j is an integer chosen randomly between 0 and a pre-defined limit. Since the number of modules in a given library is 10, this limit is also set to 10. We use five of these tag-reference generators in our set of instructions available for the programs and modules, meaning that referencing instructions will be chosen five times as frequently as any other specific instruction.

10.3.3 Updating the Library

The algorithm, Fig. 10.2, presents the modified steps of the genetic programming algorithm to accommodate the steps needed to update the library containing modules for every individual. In the algorithm, the `ind` variable is a data structure (dictionary, hashmap, or something similar) containing the main program and a set of modules called the library.

For every generation, whenever a new child genome is created, while applying genetic operators on it, its local library is also updated. Other details of this updating procedure for the library are given in the following subsections.

```
for each generation do
    for each reproduction event do
        if doing crossover then
            parent1 := Selection(pop)
            parent2 := Selection(pop)
            ind := Crossover(parent1, parent2)
        else
            ind := Selection(pop)
        end

        /* Update the library                                                    */
        ind['library'] := Mutate(ind['library'], mutation_rate_for_module) // Apply
            mutation on the modules
        ind['library'] := Extract(ind, extraction_rate) // Replace the unused
            modules with the new ones
        ind := Absorb(ind, absorption_rate) // Absorb modules in the main
            program as well as other modules

        ind['main program'] := Mutate(ind['main program'], mutation_rate_for_program)
    end
end
```

Fig. 10.2 Modified Genetic Programming Algorithm to accommodate the steps needed for updating the library

10.3.3.1 Crossover and Mutation

During the crossover of two individuals, if a part being received by the child contains tag references, the child also receives the associated modules. In case various parts from different parents contain the same tag referring to different modules, some of those tags may be mutated to accommodate all the modules being referred to by the tags. For example, assume the child genome receives two parts, and both of them contain `tagged_2`. In order to retain modules from both parts, one of the tags can be changed to some other tag not currently in use. If no other free tag is available, the module with tag 2 from one of the parts may be dropped.

The mutation methods for the program and the modules in the library can be different, but to maintain uniformity and simplicity, we will assume the same method is used for both parts. The rates of mutation, however, might be different for the main program and modules in the following ways:

1. The mutation rate for modules is less than that of the program. In extreme cases, the mutation rate can also be zero. That would mean that mutation is not applied to the modules.
2. The mutation rate for modules is the same as that of the main program.
3. The mutation rate for modules is more than that of the main program.

This difference in mutation rates can affect the dynamics of evolution. In the first case, for example, lower mutation rates essentially mean the modules are being shielded from frequent mutations so that they can evolve code segments which might be too useful for the programs to be changing frequently. This is also similar to the idea of multiple levels of evolution in hierarchical evolution [2], where modules in the lower layers evolve at a slower speed than the modules or programs in the upper levels. Keeping the mutation rate for modules same as or more than that of the program would not provide any such protection to the modules.

Note that although we discussed crossover and mutation operators in detail here to show the generality of GLEAM, in our experiments conducted in this chapter, we only use the mutation operator. The reasons for doing this will be discussed later in the chapter.

10.3.3.2 Extraction

We consider a module being *used* if it is reachable by the program, i.e., if the program contains a reference to it, or if the program contains a reference to some other module which in turn contains the reference to it, and so on. In the extraction operator, every unused module gets replaced by a new module with a certain probability. Figure 10.3 gives an example of an unused module being replaced by a code segment from the program. Where do these new modules come from? Various methods [2, 5, 15] find the new modules using a process known as Module Search or Module Identification, where different subtrees from individuals in the population are assigned fitness depending on their usefulness in the original trees, and the subtrees with better fitness than the others are chosen as new modules. The fitness of a subtree can be assessed in various ways: difference in the fitness of the individual tree with and without the subtree, difference in the fitness with the given subtree and a randomly generated subtree inserted in its place, etc. For GLEAM, new modules may come from these places:

1. generate a random code segment of a certain length
2. chose a random code segment from the current program or any other program in the population
3. choose an existing module from the library of the current program or any other program in the population
4. any other external repository

For the first two options, the lengths of new modules would also need to be determined. This can be done by sampling lengths from a discrete distribution such as Poisson distribution, Negative Binomial distribution, etc. The above-mentioned procedures for new modules is for linear genomes. For tree-based representations, a subtree of the program can serve as the module, similar to what is done in Run-transferable libraries [5] or Hierarchical Genetic Programming [2].

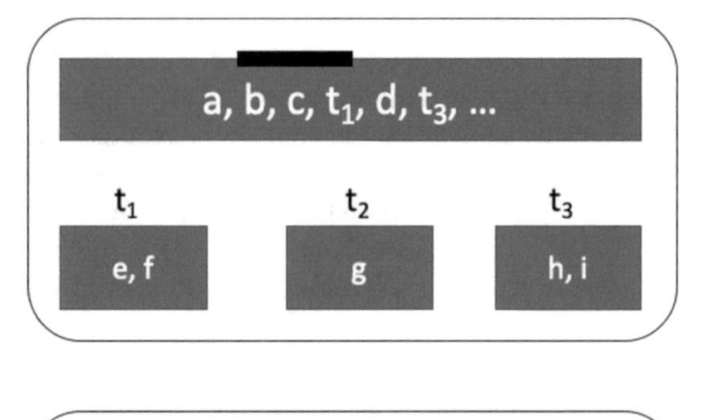

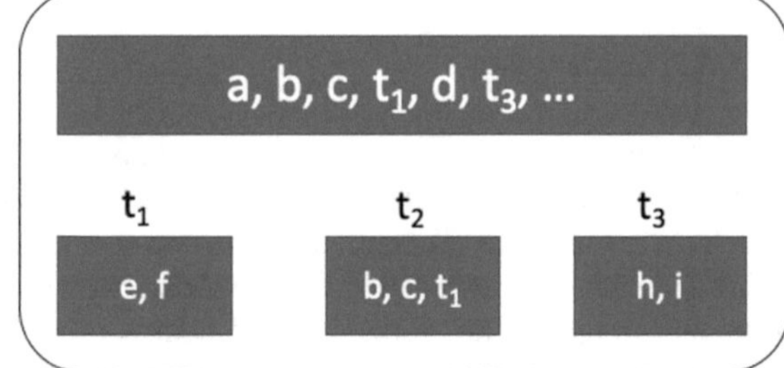

Fig. 10.3 An individual in GLEAM before and after the extraction operation. Since the tag t_2 is not being used, it is replaced with the segment bct_1 extracted from the program

10.3.3.3 Absorption

In the absorption operator, with a certain probability, every module reference, in the program as well as the modules, can get replaced by the code segment it refers to. Figure 10.4 provides an example of a module getting absorbed by the program. One reason why expanding modules might be useful for the programs is that it frees up the tag unless the same tag is referenced somewhere else in the program. The freed up tag-module association becomes unused and can be replaced by a new module. Another reason might be that once a module is expanded into the program, it can interact more freely with other instructions in the program than was possible before.

10.4 GLEAM as a Platform for Testing

Due to its flexible architecture and tunable parameters, GLEAM framework can be used as a general-purpose platform to test various aspects relating to modularity in genetic programming: how to use current modules, how to generate new ones, how to decide which old modules to replace, where do the modules actually reside, etc.

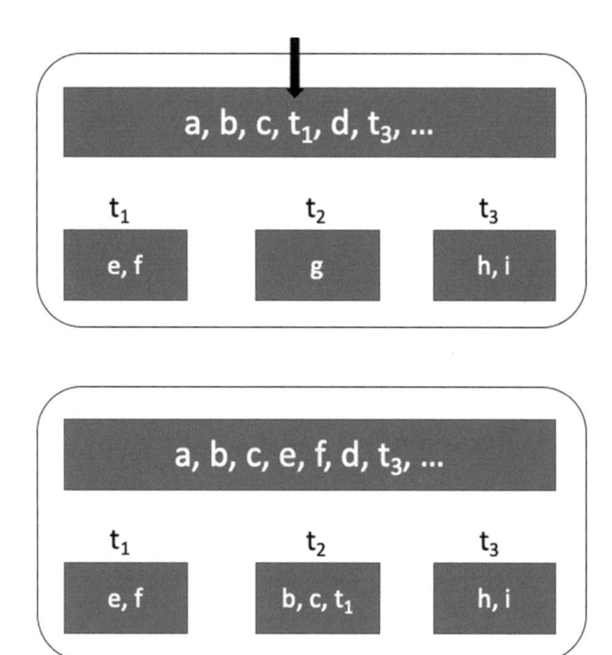

Fig. 10.4 An individual in GLEAM before and after the absorption operation. The tag t_1 gets expanded in the program

We put these methods in four categories and discuss some examples of each of them. The categories are listed in Table 10.1 and are described below:

1. **Using modules:** How do modules get referenced? The methods to do that may include referencing by an identifier, insertion of the modules directly into the program, etc.
2. **Storing modules:** Where do modules reside? They might be kept in a library local to an individual, a library common to all the individuals in the population, or a combination of both. They can also be defined as part of the program itself.
3. **Generating Modules:** Where do new modules come from? They can come from the segments of the program using them or any other program in the population, or any external repository. They can even be generated randomly from the instruction set available to the program.
4. **Replacing modules:** Which modules get replaced whenever new modules are being added? Various policies that can be adopted are: replace the ones not used by the program, replace the ones least recently used, replace the oldest ones, etc.

Table 10.1 Some of the methods that can be tested using GLEAM framework

Using modules	**Storing modules**
Reference by an identifier	In the program
Direct insertion	A library local to the program
	A library shared by all individuals
Generating modules	**Replacing modules**
From the program using the modules	Replace the ones not used by the program
From other programs	Oldest
External repository	

10.5 Experiments and Analysis

In this section, we describe some experiments conducted to test the effectiveness of GLEAM to evolve modular programs and to test various methods relating to the generation, usage, etc. of modules during evolution.

10.5.1 Experimental Set-Up

All experiments were conducted in a genetic programming system called PushGP; specifically, a version of PushGP written in Clojure, called Clojush,[1] was used. This system evolves programs in a stack-based programming language called Push [12, 14]. In this language, every data type has a dedicated stack, and during execution, instructions can take their inputs from and place their outputs on different stacks.

To test the effectiveness of GLEAM, we ran it on five problems from the General Program Synthesis Benchmark Suite [4]. These problems have been selected keeping in mind their difficulty level—state of the art [4] gives about 50% or less success rate on these problems—as well as the input and output data types. Descriptions of these problems are reproduced here:

- Last Index of Zero: Given a vector of integers, at least one of which is 0, return the index of the last occurrence of 0 in the vector.
- Count Odds: Given a vector of integers, return the number of integers that are odd, without the use of a specific even or odd instruction (but allowing instructions such as mod and quotient).
- Compare String Lengths: Given three strings $n1$, $n2$, and $n3$, return true if $length(n1) < length(n2) < length(n3)$, and false otherwise.
- Small or Large: Given an integer n, print 'small' if $n < 1000$ and 'large' if $n \geq 2000$ (and nothing if $1000 \leq n < 2000$).

[1] https://github.com/lspector/Clojush.

Table 10.2 Genetic Programming Parameters

Parameter	Value
Population size	1000
Number of generations	300
Parent selection algorithm	Lexicase
Mutation operator	UMAD
Mutation rate	0.09
Genome Representation	Plushy
Number of runs per condition	50

- Double Letters: Given a string, print the string, doubling every letter character, and tripling every exclamation point. All other non-alphabetic and non-exclamation characters should be printed a single time each.

Some of the genetic programming parameters used for the experiments are given in Table 10.2. For the problems in the benchmark suite, the best results [3] so far have been obtained with using lexicase selection (or its other variants) as selection operator, uniform mutation by addition and deletion (UMAD) as the mutation operator, and no crossover operator. This is the setting we use in our experiments as well.

10.5.2 Using GLEAM to Evolve Modular Programs

As far as parameters specific to GLEAM are concerned, we conducted some preliminary experiments with various settings, and the configuration that worked the best, in terms of the number of successes on the problems from the benchmark suite, is given in Table 10.3. In this configuration, the mutation operator used for the modules is the same as that of the program, i.e., UMAD, but the rate used for the modules is half of what is used for the program. Initial modules have the size one-tenth the size of the initial programs. Every unused module is replaced with a probability of 0.75 by a randomly chosen sequence of genes that is extracted from the program itself.

New modules are extracted in the following way. First, a number is chosen randomly between 1 and 20. This number called $len\_segment$ serves as the tentative length of the new module. Then, a random point in the program genome is chosen. Let's call it $start$. Now, the segment between $start$ and $min(start + len\_segment, len\_program)$ is extracted. After extracting the module, we check whether it has balanced parentheses. If not, we choose the first subsegment of the extracted segment with balanced parentheses as the new module.

We also allow module references to get absorbed in the program as well as in other modules with a probability of 0.1. That means one out of every ten module references would be expanded.

Table 10.3 GLEAM Parameters

Parameter	Value
Number of modules per individual	10
Mutation operator for modules	UMAD
Mutation rate for modules	half of what is used for the program
Extraction method	sequence from the program
Extraction rate	0.75
Module absorption rate	0.1

Table 10.4 Number of successes out of 50 for various configurations

Problem	W/out GLEAM	Config A	Config B	Config C	Config D	Config E
Last index of zero	29	37	34	36	33	37
Count odds	3	8	5	4	5	3
Compare string lengths	14	19	17	11	13	15
Small or large	4	4	6	4	4	6
Double letters	13	12	11	16	13	12

Table 10.4 gives the number of successes, out of 50, for various problems. The first column gives the number of successes without using GLEAM. The second column, i.e., Config A corresponds to the parameter settings given in Table 10.3. To qualify as a solution, the program must have a zero error on the training set, which was used during evolution, and also on a held-out test set, which was not used during evolution. Specifically, the programs which produced zero error on the training set were first simplified according to the procedure detailed in [4], and then were run again on the test set. Those simplified evolved programs producing zero error on the test set were termed solutions. Automatic simplification was done because it has been to improve generalization on the unseen test set.

Although the improvement in the success rate under Config A for the benchmark problems is moderate and not statistically significant (according to pairwise chi-square test with Holm correction and a 0.05 significance level), this improvement is consistent across most of the parameter settings we have tested with GLEAM. The future work will entail testing various methods available in the literature of generating new modules, storing those modules, etc. to figure out which one of them will work the best with these problems.

10.5.3 Using GLEAM as a Testing Platform

We also conducted some additional experiments to illustrate the effectiveness of GLEAM to be used as a testbed for testing various methods of generating, propagating, and using modules. To that end, we experiment with some of the conditions described in Sect. 10.4. We call the parameter settings of Table 10.3 "Config A" and compare the performance of all other settings with this one. In each of the configurations described below, only one parameter is changed while keeping everything else the same as in Config A.

1. **Config A**: The parameter setting of Table 10.3.
2. **Config B (Random Segments)**: In this configuration, new modules come from randomly generated segments and not from the program. To keep the comparison fair, we calculate the lengths of the new modules as if they are coming from the program according to the procedure described in Sect. 10.5.2. After that, the genes are generated randomly instead of getting extracted from the program.
3. **Config C (No Absorption)**: We set the absorption rate to be zero. This means there is no absorption either in the program or other modules.
4. **Config D (High Absorption)**: We set the absorption rate to 0.25. This is meant to test whether higher absorption is better or not.
5. **Config E (Modules don't mutate)**: There is no mutation at the level of modules.

The results are given in Table 10.4. Again, since the changes in the success rate are moderate and not statistically significant, we will mainly talk about some general trends rather than any concrete conclusions drawn from the data. The following trends may be observed:

1. Using segments extracted from the program itself as modules seems to be better than using randomly generated segments.
2. Some absorption is better than having no absorption or very high absorption.
3. Having mutation at the level of modules seems to work better than not mutating modules at all.

10.5.4 Modular Usage in GLEAM

Do programs evolved using GLEAM actually use modules? We answer this question using the following metric. For each solution program, we calculate the number of module references that have been absorbed in the programs at various points in its lineage. This number added to the number of module references currently in the solution program will give us the total number of modules that have been used to evolve the solution. We report this number averaged over all the solutions for a given problem. The data is presented in Table 10.5.

Since the solution programs and the individuals in their lineages do use modules, it is possible that these modules contain some useful information that was used by

Table 10.5 Average number of modules used to evolve the final solutions

Problem	Config A	Config B	Config C	Config D	Config E
Last index of Zero	3.32	2.38	1.22	3.58	4.35
Count odds	5.25	13.0	2.5	20.2	13.0
Compare string lengths	9.95	10.12	1.73	15.23	12.27
Small or large	1.75	8.5	1.75	14.0	6.33
Double letters	31.92	25.54	2.69	35.77	25.17

them. It is also possible that these modules might have helped the individuals find new pathways to solutions through the search space. Further research is needed to verify these claims.

Although the numbers presented in Table 10.5 are indicative of the fact that modules contain some useful information, these numbers do not correlate highly with the actual number of successes for a given problem. Further study will be required to understand the significance of these results, which may depend on specific aspects of problems that affect the utility of modules.

10.6 Conclusions

We presented and analyzed a general framework for evolving modular programs in genetic programming called GLEAM. Although it can be applied to any genetic programming system, in this paper, we study its effects in PushGP, a genetic programming system that evolves programs in Push programming language. We find that GLEAM improves the success rate on multiple benchmark problems. We also describe how GLEAM can be used as a platform to test various methods of generating, using, storing, and replacing modules during the evolution of modular programs. We test some of these methods experimentally as well. We also show that solutions evolved using GLEAM often use modules, indicating their usefulness during the process of evolution.

The problems on which we have tested the GLEAM framework are relatively simple problems, in the sense that the current genetic programming systems are able to solve them with a moderate success rate. For more complex problems, especially those which have not been solved by any genetic programming system so far, the utility of GLEAM remains to be seen. Future work would also investigate the advantages of using GLEAM with other forms of genetic programming, including tree-based, grammar-based, and steady-state systems, etc. More research is also needed to examine why some parameter settings are more effective than others for problems of various levels of difficulty.

Acknowledgements This material is based upon work supported by the National Science Foundation under Grant No. 1617087. Any opinions, findings, and conclusions or recommendations expressed in this publication are those of the authors and do not necessarily reflect the views of the National Science Foundation. This work was performed in part using high performance computing equipment obtained under a grant from the Collaborative R&D Fund managed by the Massachusetts Technology Collaborative.

References

1. Angeline, P.J., Pollack, J.B.: The evolutionary induction of subroutines. In: Proceedings of the 14th Annual Conference of the Cognitive Science Society, pp. 236–241. Bloomington, Indiana (1992)
2. Banzhaf, W., Banscherus, D., Dittrich, P.: Hierarchical genetic programming using local modules. In: Bar-Yam, Y., Minai, A. (eds.) Unifying Themes in Complex Systems - Proceedings of 2nd International Conference on Complex Systems, pp. 321–330. CRC Press (2018)
3. Helmuth, T., Abdelhady, A.: Benchmarking parent selection for program synthesis by genetic programming. In: Proceedings of the 2020 Genetic and Evolutionary Computation Conference Companion, pp. 237–238 (2020)
4. Helmuth, T., Spector, L.: General program synthesis benchmark suite. In: Proceedings of the 2015 Annual Conference on Genetic and Evolutionary Computation, pp. 1039–1046 (2015)
5. Keijzer, M., Ryan, C., Cattolico, M.: Run transferable libraries - learning functional bias in problem domains. In: Genetic and Evolutionary Computation Conference, pp. 531–542. Springer (2004)
6. Kelly, S., Newsted, J., Banzhaf, W., Gondro, C.: A modular memory framework for time series prediction. In: Proceedings of the 2020 Genetic and Evolutionary Computation Conference, pp. 949–957 (2020)
7. Koza, J., Bennet, F., Andre, D., Keane, M.: Genetic Programming III. Morgan Kaufmann Publishers (1999)
8. Koza, J.R.: Genetic Programming: On the Programming of Computers by Means of Natural Selection. MIT Press (1992)
9. Lalejini, A., Ofria, C.: Evolving event-driven programs with signalgp. In: Proceedings of the Genetic and Evolutionary Computation Conference, pp. 1135–1142 (2018)
10. O'Neill, M., Spector, L.: Automatic programming: the open issue? Genet Progr. Evol. Mach. **20**, 1–12 (2019)
11. Spector, L.: Evolving control structures with automatically defined macros. In: Working Notes of the AAAI Fall Symposium on Genetic Programming, pp. 99–105 (1995)
12. Spector, L., Klein, J., Keijzer, M.: The push3 execution stack and the evolution of control. In: Proceedings of the 7th Annual Conference on Genetic and Evolutionary Computation, pp. 1689–1696 (2005)
13. Spector, L., Martin, B., Harrington, K., Helmuth, T.: Tag-based modules in genetic programming. In: Proceedings of the 13th Annual Conference on Genetic and Evolutionary Computation, pp. 1419–1426 (2011)
14. Spector, L., Robinson, A.: Genetic programming and autoconstructive evolution with the push programming language. Genet Progr. Evol. Mach. **3**(1), 7–40 (2002)
15. Swafford, J.M., Hemberg, E., O'Neill, M., Brabazon, A.: Analyzing module usage in grammatical evolution. In: International Conference on Parallel Problem Solving from Nature, pp. 347–356. Springer (2012)
16. Walker, J.A., Miller, J.F.: The automatic acquisition, evolution and reuse of modules in cartesian genetic programming. IEEE Trans. Evol. Comput. **12**(4), 397–417 (2008)

Chapter 11
Evolution of the Semiconductor Industry, and the Start of X Law

Andrew N. Sloss

Abstract In this paper, we explore the use of evolutionary concepts to predict what-comes-next for the Semiconductor Industry. At its core, evolution is the transition of information. Understanding what causes the transitions paves the way to potentially creating a predictive model for the industry. Prediction is one of the essential functions of research; it is challenging to get right; it is of paramount importance when it comes to determining the next commercial objective and often depends on a single change. The most critical part of the prediction is to explore the components that form the landscape of potential outcomes. With these outcomes, we can decide what careers to take, what areas to dedicate resources towards and further out as a possible method to increase revenue. The Semiconductor Industry is a complex ecosystem, where many adjacent industries rely on its continued advancements. The human appetite to consume more data puts pressure on the industry. Consumption drives three technology vectors, namely storage, compute, and communication. Under this premise, two thoughts lead to this paper. Firstly, the *End of Moore's Law* (EoML) [33], where transistor density growth slows down over time. Either due to costs or technology constraints (thermal and energy restrictions). These factors mean that traditional iterative methods, adopted by the Semiconductor Industry, may fail to satisfy future data demands. Secondly, the quote by Leonard Adleman "*Evolution is not the story of life; it is the story of compute*" [2], where essentially evolution is used as a method to understand future advancements. Understanding a landscape and its parameterization could lead to a predictive model for the Semiconductor Industry. The plethora of future evolutionary steps available means we should probably discard focusing on EoML and shift our attention to finding the next new law for the industry. The new law is the *Start of X Law*, where *X* symbolizes a new beginning. Evolutionary principles show that co-operation and some form of altruism may be the only methods to achieve these forward steps. Future choices end up being a balancing act between conflicting ideas due to the multi-objective nature of the overall requirements.

A. N. Sloss (✉)
Arm Ltd., Seattle, Washington, USA
e-mail: Andrew.Sloss@arm.com

W. Banzhaf et al. (eds.), *Genetic Programming Theory and Practice XVIII*,
Genetic and Evolutionary Computation,
https://doi.org/10.1007/978-981-16-8113-4_11

11.1 Introduction

Similar to many industries, the Semiconductor Industry is and has been, driven by immediate short-term goals, e.g., 65 nm, 28 nm, and 14 nm manufacturing processes. These short-term goals are what we call the *Hamster Wheel Effect*, see Fig. 11.1. As time progresses, these immediate goals get harder and harder to achieve, as the *degrees of evolvability* reduces. Degrees of evolvability reflects what changes are possible, and this is a property of lineage (what came before) [7]. As-in, history determines what options are available for future advancements.

Every little advancement overcomes a hurdle, which in turn allows for the exploration of new nano-scale worlds. A good recent example is *Extreme Ultraviolet Lithography* (EUV) [24]. EUV operates at a wavelength of 13.5 nm, which allows for stable manufacturing at 5 nm, 3 nm scales, and possibly beyond.

At some point, increments are exhausted, and there are no further seams left to explore, leaving technology-jumps as the only option. Technology-jumps are significant shifts in the industry. Compared to increments, these jumps remove constraints that had impeded progress, but normally introduce new limitations, i.e., jumps significantly change the landscape.

As we gain a greater understanding of biological evolution, we find that it is driven by opportunistic changes over a much longer timescale. One of the most significant jumps in biology was the incorporation of mitochondria and *plastids* for energy production and photosynthesis, forming the first ancient *eukaryotic* cell. The combination of power-production, replication, and computation into a single unit allowed for the existence of multi-cellular organisms. These organisms set the stage

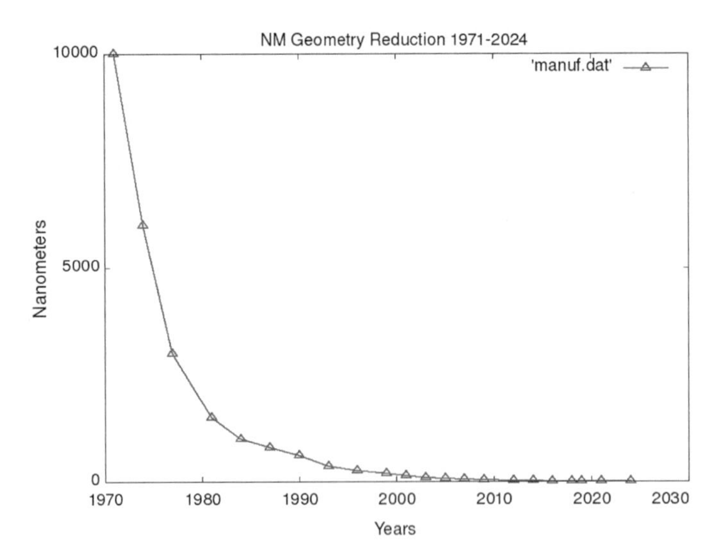

Fig. 11.1 Silicon manufacturing process scale since 1971

for what is commonly called the *Cambrian Explosion*, a short period when a dramatic acceleration of diversity occurred within the animal kingdom. In other words, a jump to a higher level caused a rapid expansion of options and variety.

Both the Semiconductor Industry and biology may seem poles apart and focus on fundamentally different substrates. But both systems are complicated, deal with roughly the same geometries, and have the challenges of handling vast interactions (spatial and temporal) to achieve useful work, i.e., a complex system. A complex system includes the history, as-in the lineage. Thurner et al. described a complex system as a *co-evolving network* of entities [32]. Where co-evolving refers to two or more groups affecting each other's evolutionary progression.

Before continuing further, let us define what is meant by evolution. Evolution is a continuous process cycling through replication, selection, and variation. A single cycle produces a population of potential solutions. A complete cycle is called a *generation*. Evolution explores adjacent possibilities either locally and/or globally [17, 32]. The continuous process allows the solutions to adapt dynamically to changing environments, and the entities within the population have active fitness values. For engineering, these values represent the distance to a desired goal or goals.

Novelty, or uniqueness, is predominately controlled by the level of *diversity* within the population. If a new idea is required then a more diverse population is more likely to succeed. A diverse population increases the likelihood of something new emerging. This find assumes there is something new that has not been discovered. If the population already includes every possibility in the space, nothing new can emerge; in such a case, a less diverse population increases the likelihood of something new emerging that is not a new possibility, but a new actuality in the population.

Rapid evolutionary advancements occur when there is significant environmental pressure. The pressure either causes the population to converge to a new characteristics or it remains diverse. If convergence occurs too early in the process a potential "better" solution could be missed. Biological pressure points have come from significant changes in the environment. Either by a reduction or sudden increase in the availability of a resource, e.g., oxygen. By comparison, data consumption can be seen as the main pressure point for the Semiconductor Industry. Where projected data demand [9, 28] outstrip future resources and capabilities. In other words, the current silicon technology cannot keep up with the coming data processing demands.

Today, taking the Semiconductor Industry as a complex evolving population of systems (a population of populations), it has relatively little diversity and relies entirely on the exploitation of current skill-sets. Those skill-sets focus on digital logic etched onto silicon substrates. There are analog components, but these are kept relatively small by comparison to the digital pieces. The industry has converged. If there is a desire to escape the convergence then some form of technology-jump has to occur.

Technology-jumps require some form of *evolutionary creativity*. Meaning an inspirational change from the normal development flow has to occur. *Margaret Boden* [6], broke down creativity into three useful mechanisms, namely *exploration*—play within the rules, *combination*—apply one set of rules to another domain, or *transformation*—rewriting the rules by removing a critical constraint. Exploration creativity means being risk averse, and at the other end of the scale, transformation creativity means taking the highest risk. We will use these definitions to categorize future technologies.

It should be stressed that true biological evolution is full of exceptions and controversies. For this paper, we frame *change* using only the high-level rules. Keep in mind that there are plenty of exceptions to those rules in the natural world.

11.2 Human Knowledge Constraint

To achieve a technology-jump, it is worth talking about knowledge and, more precisely, the dissemination of knowledge. Knowledge is one of the biggest constraints for a any technology-jump. Knowledge provides a distinction between entities (be that at the organizational or individual level). Knowledge is one of the elements that forces diversity within an industry. It is also a component for convergence and a potential barrier for forward progress. Valuable knowledge is created and held by a few, as time passes, it naturally disseminates through a variety of mechanisms, namely reverse engineering, personnel movement, teaching, and literature. For fun, let us put forward the general *Law of Disseminating Knowledge*.

The law states that human knowledge will disseminate at a rate inversely proportional to the value, i.e., low-value knowledge circulates fast, and high-value knowledge moves much slower. The slowness is due to the difficulty and effort required for its acquisition. Significant value tends to be latent, as-in value comes over time as its importance occurs to more people. Once valued knowledge is recognized as essential, and goes beyond a popularity threshold, restrictive control measures reduce the overall distribution, i.e., the value goes up. As an extreme example, *Atomic Bomb* knowledge is of high-value and requires severe restrictions on the distribution.

For the Semiconductor Industry, transformation change requires that the highest value knowledge must disseminate. For example, the success of Quantum Computing requires knowledge sharing beyond a few people, which requires overcoming some natural barriers such as training and lack of skilled practitioners. Another feature of this law is that low-value ideas disseminate faster, potentially undermining high-value knowledge. The Law of Disseminating Knowledge acts as a physical constraint on evolutionary creativity—if more than one organization has to participate.

11.3 Evolutionary Concepts

11.3.1 What Evolutionary Components Can Be Applied to the Semiconductor Industry?

Firstly, let us break the question down into the three creativity methods discussed previously. Where exploration creativity are the non-exotic options (local search, mutations), the combination creativity being a mixture (global search: crossover), and transformation creativity being the exotic options (local and global search, mutation + crossover). Keep in mind that evolution is dynamic and adaptive; there are continuous generational changes. Generation in this context is about satisfying the next data consumption goals.

1. **Exploration evolution**, *near term* the Semiconductor Industry revolves around the fundamental parameters we have today, namely general-purpose compute, wafer-size, design scale, synchronization, communication, density/cost, knowledge representation, tool abstraction, degrees of specialization, power constraints, new simulation technologies, performance and physical topologies (e.g., 2D or 3D). These parameters are not mutually exclusive and remain within the constraint of a silicon substrate. What is evolving from these parameters? Accelerators, Domain Specific Architectures [15], Analog Circuits, Asynchronous design, Wafer-scale chips [12], modular Chiplets [29], Compute-in-Memory [19], abstract design tools [5], and lastly, EUV advancements [24].
2. **Combination evolution**, *medium-term* relies on combining the rules of one domain into another. Examples include Machine Learning → (rules applied to) computer architecture design, cloud computing (weather prediction models of design) → simulation, adaptive Evolutionary Algorithms → Deep Neural Network hardware [13], software → hardware (FPGAs) [27], new advanced benchmarks → hardware [16], Computer Science → Biology [18], Quantum Computing → Machine Learning [21]. Probably the most significant combination jump would be to change the substrate (e.g., transition to Gallium Nitride GaN [8]) while keeping the original parameters outlined in (I) the same.
3. **Transformation evolution**, *long-term* as applied, are extremely high-risk changes that are not guaranteed to be successful, namely Quantum Computing [1], Probabilistic Computing [34], Reversible Computing [25], Self-Assembly [11], DNA Computing [35], Biological Computing [31], Neuromorphic Computing [14] and lastly Optical/Photonics Computing [3]. Changes in the environment can force the existence of high-risk alternatives, e.g., Probabilistic Computing may come about naturally as the number of error-correcting bits increases. These evolutionary changes may remain specialized, as with Jet Engines, specialized but essential for a specific purpose.

Each one of components listed above has distinct advantages and disadvantages, with differing levels of risk. Evolution shows us that there is plenty of options for continued forward momentum, even for current technology, putting credence to the view that a more optimistic message for the future is required. It is also worth stating that change comes with an associated cost.

11.3.2 What Else Does Evolution, and Economic Models Tell Us?

Long-term sustainable systems require a balance between co-operative and competitive behavior [10, 20]. Where too much co-operative action results in stagnation and too much competitiveness results in brittleness [10], see Fig. 11.2. The window of viability is where successful stable systems live for optimal long-term sustainability and *homeostasis* (moving towards stable equilibrium between interdependent elements).

Evolutionary change occurs at different levels of the stack (whether it is biological or computing), where many changes can coincide. A transformation change might ultimately involve forming a higher level of abstraction (the shift to an *Intellectual Property* model being an example of a transformational change). Mistakes and risk-taking are critical for higher-order advancements. It is also worth mentioning co-evolution, where different species influence each other in the evolutionary process [7], e.g., in Quantum Computing the influence of Google, IBM, and Microsoft have with each other, can be thought of as a co-evolutionary process.

A healthy and productive population is one that is diverse and remains diverse. A community that goes towards a mono-culture, even if thriving, does not necessarily continue to advance in the long-term. Mono-culture is ripe for disruption. As-in it remains at a lower part of the system. Bacteria being an excellent example in biology, vital as they may be, bacteria by design does not move-up the abstract scale (macro-evolution) and has remained at the same level. A bacteria today will

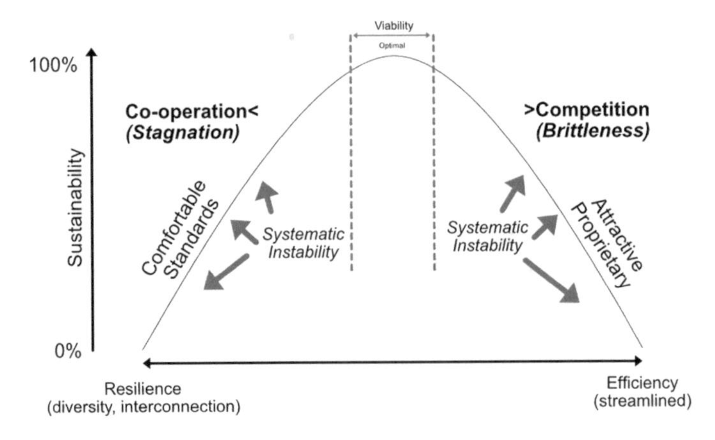

Fig. 11.2 Long Term System Sustainability [10, 20]

resemble its ancestor from 1000 generations ago, their only method of evolution, or more precisely micro-evolution, is via knowledge dissemination through a technique called *Horizontal Gene Transfer* [26] (side-note: this is the mechanism that causes antibiotics to lose their effect). Bacteria are super successful at a much lower level, but never achieve ascension to a higher level of abstraction. Success occurs at a fix level.

11.3.3 How Can Ascension Occur?

In biology, ascension to a higher-level uses two mechanisms, namely *egalitarian* or *fraternal* transition [7]. Egalitarian transition occurs when different 'skilled' lower-level entities come together to create a higher-level entity. As an example, the compartmentalization of the molecules in proto-cells (mitochondria within a eukaryotic cell). It increases individual sophistication. By comparison, the fraternal transition is where 'like' entities come together to create a higher-level entity. The best example of this is the forming of multi-cellularity (eusociality in animals). It increases system-wide sophistication. Both mechanisms adhere to the phrase "*the whole is greater than the sum of its parts*". After a high-level entity is created a shift occurs towards resilience and stability. In other words, a form of knowledge dissemination is required for it to remain relevant. The paradox here is that the knowledge needed for ascension is most likely high value, which is harder to share, but sharing is paramount to reach a higher-level. Emergent behavior in a populations at some point has to share information.

11.3.4 What About the Human Element?

Lastly, there is the human element or what we will call *Social Evolution*. Where each generation has a style of engineering that differs from the previous one. For humans, an engineering generation is about 25-years [22]. These styles tend to be transformational by nature, in that they change the abstract model from generation to generation. The groups that resist these changes are the ones that have excellent co-operation, i.e., the highest associated cost of change. The drivers for transformational changes are directly related to the active problem-domains.

For the Semiconductor Industry, data consumption and efficiency has been the critical problem-domains. The transformational shift occurring at the moment, within the industry, is the shift towards *Machine Learning* (correlation). Where the transition is from *rule-based systems* to systems that use *outcome driven data models*.

Generally, if there is a requirement to produce more of "something" (e.g. Machine Learning), then that "something" is made more abstract and cheaper to handle. In other words, there is pressure to move away from specialized skills to produce more efficient scaling tools that incorporate the *learning's* of the craftspeople in that area.

Pushing the majority to a higher abstract level. For the the Semiconductor Industry a good example of social evolution is the appearance of the *Chisel* tool [5]. The tool allows a broader group of people to architect processors—lowering the barrier for entry and scaling potential solutions.

11.4 Final Discussion and Thoughts

Up-to-now we have not spent any time on the current, or the exploitative, options available today. Exploitative in the sense that the industry continues without evolutionary creativity. With the current technology, the analogy between the Semiconductor Industry + Silicon and the Automotive Industry + Internal Combustion Engine is an exciting area to explore. The Internal Combustion Engine first appeared in 1859, some 160 years ago [4], and for the most part, has remained the same. Yes, today, there are potentially better solutions threatening to displace the Internal Combustion Engine, but it still remains dominant.

Today's silicon designs may follow the same path, with many future years of continued success without requiring the historical increases in transistor density. This is because a balance has been reached between co-operative and competitive behaviors (a sustained equilibrium point). Unfortunately, if a technology-jump is required, due to exponential data consumption pressures, then the industry must take a significant risk. That risk requires potentially shifting towards an uncomfortable solution, making the sector inherently unstable. Future shifts will create new organizational hierarchies, culminating in the emergence of new winners and losers.

What are the evolutionary variables and constants?

In a predictive model, the control elements supervise how change occurs. Below are the elements that could play a role in an evolutionary predictive model for the Semiconductor Industry.

a **Horizontal/Vertical movement**: Horizontal is about staying at the same level and hopefully improving and optimizing various systems. Effectively exploitation of public knowledge. By comparison, vertical movement is about ascension to a higher level. Vertical change involves exchanging some low-level autonomy (a form of altruism) with high-level functionality, including some evolutionary creativity (see c.). Higher-level refers to a more abstract view of a specific problem, e.g., seeing the idea rather than the implementations details.

b **Co-operative/Competitive behavior**: Co-operative is about working together to solve a common problem. The main disadvantage is stagnation and a lack of nimbleness. It involves reducing the barriers for high-value knowledge transfer, whereas, by comparison, competitiveness is about selfishness (in the biological sense) and attempting to get advantages by moving ahead without working with others. It is the opposite of co-operative behavior since it is a non-sharing approach that ultimately causes brittleness [10].

c **Exploratory/combination/transformation creativity**: Creativity is about discovering new knowledge. Exploration creativity is about following-the-rules and exploring all the points within a known boundary box. When the variables and rules are already known. The edge of the box tends to result in most the creativity. Combination creativity involves taking one set of rules applied to another domain. It involves transferring knowledge gained in one area and applying it to another. And finally, transformation creativity is about changing the rules and removing a constraint.

d **Egalitarian/fraternal transition**: is about ascending to a high-level of abstractions. Egalitarian transition is grouping functionally different lower-level entities to build a new heterogeneous individual with more capability. By comparison, the fraternal transition is about gathering similar lower-level bodies together. An organizational method to attain a higher level.

e **Knowledge dissemination**: is a associated with knowledge sharing. High-value knowledge tends to move more slowly since it requires people at the same level to understand solving identical problems. Low-value knowledge disseminates fast since the barriers of transfer are effectively nonexistent. Low-value knowledge can appear to outmaneuver high value using speed of dissemination.

f **Boolean/Statistical domain**: deal with the levels of modeling. A boolean domain is a precise domain with distinct rules. By comparison, the statistical area moves closer to the organization of nature, i.e., stochastic. Where the answers are between 0 and 1 and come with a 'certainty' component.

g **Social Evolution**: is a variable where each generation of engineers want to do something new and different. They go into the workforce with various tools and different problems to solve.

h **Vision, competition, and revenue styles**: these are variables that drive businesses. Vision is about moving towards some form of an idealistic goal. Competition is about the reaction to other players in the environment. Finally, revenue is where everything should end up, some profitable outcome.

i **Diversity**: controls the degree of novelty within the population.

j **Individuals**: are the entities that have collected different biases.

What are the technology entities to play with?

Figure 11.3 below is an attempt to show the technology entities. The diagram includes both exploratory and transformation entities, along with the separation of Boolean and statistical domains (i.e., shifting from synthetic to nature). Note, this is very much a subset and is somewhat arbitrary, and is provided as an example of a possible future technology landscape. The horizontal axis represents the difficulty of change, as-in the left side are increments (less difficult), and the right side are technology-jumps (more difficult). The vertical axis represents the abstraction level, the bottom half

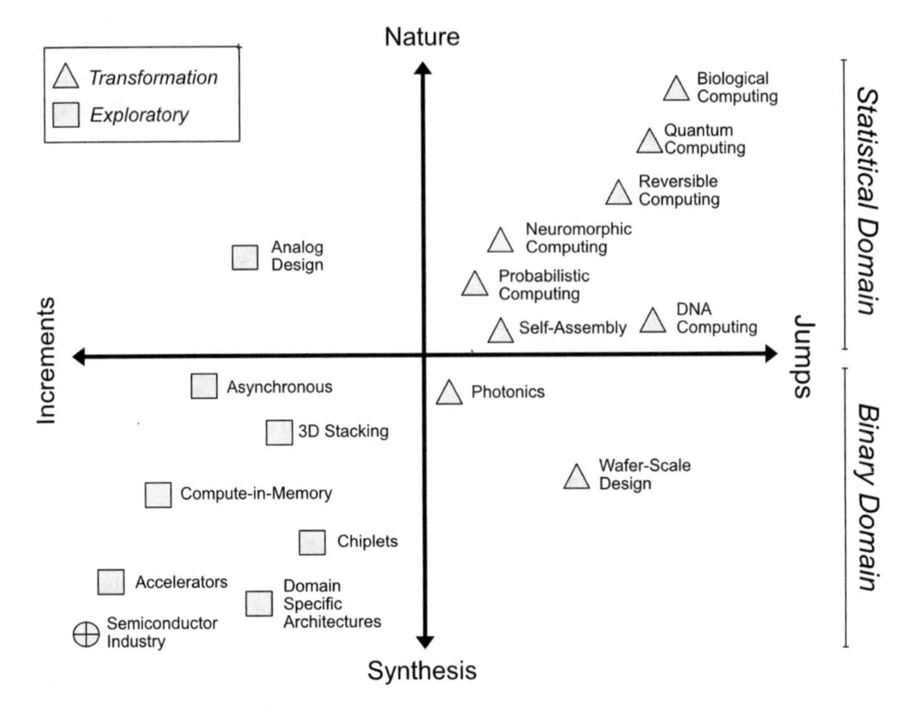

Fig. 11.3 Future Technology Landscape beyond Moore's Law

shows binary representation (traditional digital systems), and the top half represents more of the natural world, i.e., stochastic and statistical.

What are the drivers?

Healthy, expanding industries tend to follow an exponential growth curve. Historically, Moore's Law [23] has been used as the target for the Semiconductor Industry. There are a set of constraints that generally come along with growth, namely power consumption, environmental-costs, and resource usage. For the Semiconductor Industry, growth has been more critical because of all the adjacent industries that rely on continued advancements.

In economics, the *Jevons paradox* [30] comes into effect. Paraphrasing, "*as technology progresses, the efficiency of unit resource-use improves, and subsequently, the rate of system resource-use rises due to increased demand*". In other words, as we provide better methods of data consumption, we create a cycle that expects evermore capabilities. To continue satisfying those demands, a significant technology-jump must occur. The Internal Combustion Engine reached its upper limits many years ago, with impressive improvements in both pollution reduction and fuel efficiency. The question for the Semiconductor Industry is whether those same limits have been reached with silicon. Are the demands for data consumption outstripping future technological advancements? If so, then a technology-jump, or jumps, must occur or the Semiconductor Industry risks limiting growth to incremental improvements or worse an external disruption.

Examples of exponential growth, once a technology-jump has occurred, are the Industrial Revolution ∵ (because of) steam power, Cambrian Explosion ∵ multi-cellular organisms, and the Semiconductor Industry ∵ continued reduction in transistor size. All showing rapid growth due to high-value knowledge dissemination. Ultimately culminating in a higher level of understanding and opportunity.

11.4.1 What are the Mechanisms for Continued Exponential Growth?

From an evolutionary standpoint, one mechanism for growth is ascension to a higher abstract level. This is possible if the limits of hierarchical complexity have not been reached. Ascension can emerge through co-operation and co-evolution; in other words, the occurrence of high-value knowledge dissemination. This emergence gives the industry an excellent chance to increase resilience, longevity, compatibility, support, knowledge sharing, and finally, the ultimate goal of revenue growth. Transition to a higher-level involves either building an entity with unique skills or focusing on an ecosystem (made-up of companies, nonprofits, and governments). The difficulty of any transition is deciding upon division-of-labor (i.e., who does what), continuity towards a long-term goal (i.e., consistency), and the handling of defections. If all three challenges are satisfied then a technology-jump, or jumps, can occur.

11.5 Conclusion

This paper was a discussion about revolution over iteration, and the associated elements required to create a predictive model. The industry has three choices. The first, organizations within the Semiconductor Industry, remain *isolated, and exhibit competitive behaviors*. This is fine for iterative short-term growth but tends to fail on long-term sustainability i.e., ripe for disruption. Second, organizations exhibit the *same level of co-operative behavior*. This is more like the Internal Combustion Engine or Bacteria. Relies on an iterative approach. Information is shared via Horizontal Knowledge Transfer (Horizon Gene Transfer in biology) and occurs without ascension. Again, similar to the first choice it may keep the industry-relevant, but not at the edge of computation. Third and final thought, the industry attempts to *ascend to a higher-level using some form of co-operative behavior and altruism*. A revolutionary high-risk approach. High-risk has the advantage of creating long-term sustainability at the edge of computation.

This paper does not cover the secondary factors (i.e., byproducts) of risk-taking. There are no spatial or temporal links (connecting triggers), so no discussion on lineage, induction, or causality. Also, the industry is treated as a single biosphere, with no regional differences. Every other factor is considered diversity in the broadest sense. No discussion of software as an essential evolving element in the environment.

The next step is to build an evolutionary model of the Semiconductor Industry, with the necessary variables, to explore the potential different futures. In the hope of eventually defining X Law.

Acknowledgements Each paper is only as good as the reviewers. I wish to personally thank the following people: Andrew Loats, Lee Smith, Paul Gleichauf, Greg Yeric, Karl Fezer, Joseph Fernando, Tim Street, Stuart Card, and Gary Carpenter for their participation and thoughts. Lastly, I would like to specially thank Stephen Freeland for a great talk, and initial strange discussion, on the idea of mapping the Semiconductor Industry to Evolutionary Biology.

References

1. Abigail Beall, M.R.: What are quantum computers and how do they work? WIRED explains. https://www.wired.co.uk/article/quantum-computing-explained (2018). Accessed 10 Feb 2020
2. Adleman, L.: Genes, memes, cenes. In: Genes, Memes, Cenes. Presented at the DNA 25 Compute Conference, Seattle, WA. http://misl.cs.washington.edu/events/dna25/program.html#adleman (2019)
3. American Institute of Physics: Bypassing Moore's law with high-speed photonic computers. https://scitechdaily.com/bypassing-moores-law-with-high-speed-photonic-computers/ (2019). Accessed 10 Feb 2020
4. American Society of Mechanical Engineers. Internal Combustion Engine Division. Technical Conference, A.S., Somerscales, E., Zagotta, A., American Society of Mechanical Engineers. Internal Combustion Engine Division, A.S.: History of the Internal Combustion Engine: Presented at the Eleventh Annual Fall Technical Conference of the ASME Internal Combustion Engine Division, Dearborn, Michigan, October 15–18, 1989. ICE (Series). American Society of Mechanical Engineers (1989)
5. Bachrach, J.: Chisel accelerating hardware design. https://riscv.org/wp-content/uploads/2015/01/riscv-chisel-tutorial-bootcamp-jan2015.pdf (2015). Accessed 9 Feb 2020
6. Boden, M.A.: Creative Mind: Myths and Mechanisms, 2nd edn. Routledge, USA (2003)
7. Calcott, B., Sterelny, K., McShea, D., Simpson, C., Okasha, S., Godfrey-Smith, P., Lyon, P., Kerr, B., Nahum, J., Rainey, P., et al.: The Major Transitions in Evolution Revisited. Vienna Series in Theoretical Biology. MIT Press, Cambridge (2011)
8. Chen, A.: Gallium nitride is the silicon of the future. https://www.theverge.com/2018/11/1/18051974/gallium-nitride-anker-material-silico (2019). Accessed 10 Feb 2020
9. Cisco: cisco visual networking index: forecast and trends, 2017–2022 White Paper. https://www.cisco.com/c/en/us/solutions/collateral/service-provider/visual-networking-index-vni/white-paper-c11-741490.html (2019). Accessed 6 Jan 2020
10. Clippinger, J., Bollier, D.: From bitcoin to burning man and beyond: the quest for identity and autonomy in a digital society. ID3 (2014)
11. et al., J.F.: Self-assembly. https://www.sciencedirect.com/topics/materials-science/self-assembly (2018). Accessed 10 Feb 2020
12. Feldman, A.: Cerebras Wafer Scale Engine: why we need big chips for deep learning. https://www.cerebras.net/cerebras-wafer-scale-engine-why-we-need-big-chips-for-deep-learning/ (2019). Accessed 6 Jan 2020

13. Frolov, S.: Neuroevolution: a primer on evolving artificial neural networks. https://www.inovex.de/blog/neuroevolution/ (2018). Accessed 9 Feb 2020
14. Furber, S.: Large-scale neuromorphic computing systems. J. Neural Eng. **13**(5), 051,001 (2016)
15. Hennessy, J.L., Patterson, D.A.: A new golden age for computer architecture. https://cacm.acm.org/magazines/2019/2/234352-a-new-golden-age-for-computer-architecture/abstract (2019). Accessed 9 Feb 2020
16. Industry: MLPerf: fair and useful benchmarks for measuring training and inference performance of ML hardware, software, and services. https://mlperf.org (2018). Accessed 9 Feb 2020
17. Kauffman, S.A.: The Origins of Order Self-Organization and Selection in Evolution. Oxford University Press, Oxford (1993)
18. Kriegman, S., Blackiston, D., Levin, M., Bongard, J.: A scalable pipeline for designing reconfigurable organisms. Proc. Nat. Acad. Sci. **117**(4), 1853–1859 (2020)
19. Lapedus, M.: In-memory vs. near-memory computing. https://semiengineering.com/in-memory-vs-near-memory-computing/ (2019). Accessed 9 Feb 2020
20. Lietaer, B., Ulanowicz, R.E., Goerner, S.J., McLaren, N.: Is our monetary structure a systemic cause for financial instability? Evidence and remedies from nature. J. Futures Stud. **14**(3), 89–108 (2010)
21. Louriz, R.: Highlighting quantum computing for machine learning. https://medium.com/meetech/highlighting-quantum-computing-for-machine-learning-1f1abd41cb59 (2019). Accessed 9 Feb 2020
22. McCrindle, M.: The ABC of XYZ: Understanding the Global Generations. University of New South Wales Press, Sydney (2009)
23. Moore, G.E.: Cramming more components onto integrated circuits. Electronics **38**(8) (1965)
24. Moore, K.: Euv lithography finally ready for chip manufacturing. IEEE Spectrum **5** (2018)
25. Perumalla, K.S.: Introduction to Reversible Computing, 1st edn. Chapman & Hall/CRC, Boca Raton (2013)
26. Quammen, D.: The Tangled Tree: A Radical New History of Life. HarperCollins Publishers, New York (2018)
27. Research, M.: Project catapult. https://www.microsoft.com/en-us/research/project/project-catapult/ (2018). Accessed 9 Feb 2020
28. Seagate: DataAge 2025, the digitization of the world. https://www.seagate.com/our-story/data-age-2025/ (2020). Accessed 6 Jan 2020
29. Simonite, T.: To keep pace with Moore's Law, chipmakers turn to 'Chiplets'. https://www.wired.com/story/keep-pace-moores-law-chipmakers-turn-chiplets/ (2018). Accessed 9 Feb 2020
30. Sorrell, S.: Exploring Jevons' Paradox, pp. 136–164. Palgrave Macmillan UK, London (2009)
31. Templeton, G.: How MIT's new biological "computer" works, and what it could do in the future. https://www.extremetech.com/extreme/232190-how-mits-new-biological-computer-works-and-what-it-could-do-in-the-future (2016). Accessed 10 Feb 2020
32. Thurner, S., Klimek, P., Hanel, R.: Introduction to the Theory of Complex Systems. Oxford University Press, Oxford (2018)
33. Track, E., Forbes, N., Strawn, G.: The end of Moore's law. Comput. Sci. & Eng. **19**, 4–6 (2017)
34. University of Konstanz: A step towards probabilistic computing. https://www.sciencedaily.com/releases/2019/05/190514115833.htm (2019). Accessed 10 Feb 2020
35. Watada, J.: Dna computing and its application. In: Computational Intelligence: A Compendium, pp. 1065–1089. Springer (2008)

Index

A
Action program, 2
  multi-action program, 6
Activity dependence, 166
Ascension, 203
Automated program repair, 46

B
Bees algorithm, 117
Benchmarking, 8, 84

C
Cache, 52, 161
Cambrian explosion, 199
Causality, 71
Classification, 166
Co-evolution, 202
Competition, 89, 114, 205
Context-free grammar, 48
Convergence
  phenotypic, 150, 199
Crossover
  asymmetry of GP subtree crossover, 151
  fatherless crossover, 158
  unbiased subtree crossover, 150

D
Data balancing, 133, 141
Deep learning, 2, 109, 165
  with genetic programming, 109
Diagnostics
  exploration diagnostics, 104
  selection scheme diagnostics, 104
Discriminant analysis, 113
Diversity, 2, 53, 63, 88, 139, 199
  phenotypic, 64
  phenotypic diversity, 89
  phylogenetic, 64
  phylogenetic diversity, 84

E
Eco-EA, 66
Efficiency, 84, 114, 129, 155, 203
Ensembles, 133, 138, 139
Exploration diagnostic, 67
Exponential growth, 206

F
Feedback loop, 71
Fitness
  predicting evaluation time of, 160
Fitness sharing, 66

G
General artificial intelligence, 165
Genetic learning, 182
Genetic programming
  BalancedGP, 133, 137
  grammar-based vectorial GP, 22
  networked runs genetic programming, 109
  OrdinalGP, 134, 137
  PushGP, 52, 102, 190
  template-constrained genetic programming, 45, 109
  vectorial GP, 22

W. Banzhaf et al. (eds.), *Genetic Programming Theory and Practice XVIII*, Genetic and Evolutionary Computation,
https://doi.org/10.1007/978-981-16-8113-4

Grammar-guided, 22
Graph, 28, 111, 183
Growing neural networks, 111, 168

H
High performance, 143, 195
Homeostatis, 172
Horizontal gene transfer, 203

I
Inefficient threads
  avoiding, 143
  causes, 143
  measurement, 143
  prediction, 143
Information loss, 33, 40
Inplace crossover, 143
  shuffle, 143
  speedup, 143
Intellectual property, 202

L
Lexicase selection, 65, 66, 83, 191
  cohort lexicase selection, 83
  down-sampled lexicase selection, 83
  epsilon lexicase selection, 83
  novelty lexicase selection, 83
Linear genetic programming, 3, 7, 18, 69, 184
Liquid types, 50, 51

M
Memory bandwidth, 143
Memory use
  minimising, 143
Metrics, 70
Mitochondria, 203
Modular, 167, 194
Modularity, 2, 7, 69, 181, 194
Moore's Law, 197, 206

N
Novelty, 90, 199

P
Panmictic, 146, 150
Parent selection, 65, 83, 191
Pareto tournament, 131
Partially observable, 1
Population diversity, 2, 66, 97, 199
Population initialization, 2, 5, 12, 55, 90, 174
Predicting success based on diversity, 63
Program
  dendrite program, 168
  evolving modular program, 182
  neuron program, 176
  program graph, 2, 183
  program representation, 46
  program synthesis, 47, 52, 84
  program synthesis benchmark suite, 190
programming languages, 48
Program synthesis, 47, 84, 190
  program synthesis benchmark suite, 190

R
Rampant mutation, 2
Reinforcement learning, 2, 17, 176
Resilience, 203, 207

S
Selection
  offspring selection, 34
  selection pressure, 34, 157, 161
Semantic constraints, 48
Semiconductor industry, 197
SMT solvers, 48, 58
Social evolution, 203, 205
Strongly-typed, 25
Sustainability, 202
Symbolic regression, 24, 30, 87, 88, 115, 116

T
Tags, 183
Tangled program graphs, 2, 183
Team, 3, 183
Tournament selection, 65, 85, 90, 91, 103, 130, 143, 145, 150, 161
Tree-based GP, 26, 28
Tree depth, 57
Type-aware, 50

Printed in the United States
by Baker & Taylor Publisher Services